1982 Supplement to
# DNA Replication

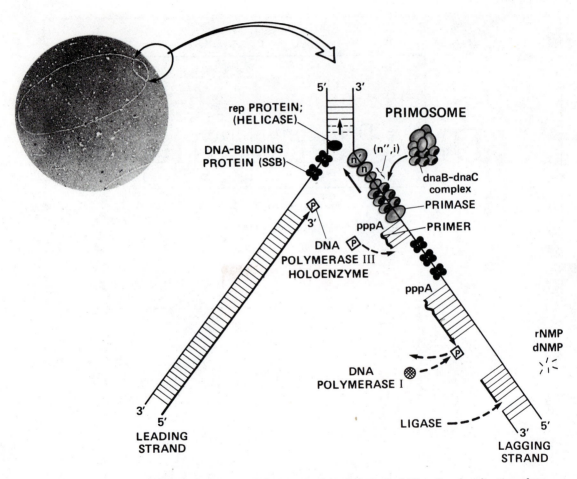

Scheme for enzymes operating at one of the forks in the bidirectional replication of an *E. coli* chromosome or of a plasmid (oriC) employing the unique origin of the *E. coli* chromosome. Replication is continuous for one strand (leading) and discontinuous for the other (lagging). Processivity (lack of dissociation) of the DNA polymerase III holoenzyme in continuous chain elongation and of the primosome in short RNA primer formation for discontinuous syntheses assures rapid replication (see Section S11–14).

# 1982 Supplement to
# DNA Replication

### Arthur Kornberg
STANFORD UNIVERSITY

## W. H. FREEMAN AND COMPANY
New York   San Francisco

**Cover:** A looped rolling circle of the duplex replicative form of phage φX174 (based on an electron micrograph provided by Dr. Jack Griffith). (Design by Marjorie Spiegelman.)

Project Editor: Patricia Brewer; Production Coordinator: Linda Jupiter; Artist: Charlene Levering; Compositor: Typothetae; Printer and Binder: R. R. Donnelley & Sons Company.

**Library of Congress Cataloging in Publication Data**

Kornberg, Arthur, 1918–
  1982 supplement to DNA replication.

  Includes indexes.
  1. Deoxyribonucleic acid synthesis. I. Kornberg,
Arthur, 1918– . DNA replication. II. Title.
[DNLM: 1. DNA replication. 2. DNA—Biosynthesis.
QU 58 K84d 1980 Suppl.]
QP624.K66 Suppl.     574.87′32     82–5117
ISBN 0-7167-1410-8 (pbk.)          AACR2

Printed in the United States of America

3 4 5 6 7 8 9  DO  1 0 8 9 8 7 6 5 4 3

# Contents

# Preface

*DNA Replication* (1980) embraces the biochemical, genetic, and physiological aspects of the process and the numerous DNA transactions that determine the structure and function of genetic material. The text and figures were designed to orient and inform students beginning their study of DNA replication as well as professionals already working on the subject. The biochemical emphasis is evident in the devotion of half the book to key enzymes and proteins (polymerases, ligases, nucleases, binding proteins) and the remainder to the cellular mechanisms in which they are employed (replication, repair, recombination, viral life cycles).

Progress in DNA replication has continued at a pace that demands a major addendum after only two years. This 1982 Supplement cites items and key references that alert the reader to important developments from June 1979 to January 1982. It does not repeat the essential facts, figures, ideas, and references in *DNA Replication*. This book is therefore a biennial news summary and not a compendium of *DNA Replication*. The surprising size of this Supplement stems both from the volume of advances in DNA structure, enzymology, and functions and from a wish to recount these advances intelligibly (with numerous pictures and references) rather than telegraphically. In addition, each of the seventeen chapters opens with an abstract, highlights selected with considerable editorial license.

In many special areas I have relied on the knowledge and enthusiastic help of colleagues. Among them, I want to cite especially Roger Kornberg for DNA structure, Mary Ellen Jones and Thomas Traut for biosynthesis of nucleotides, Peter Reichard for the latter and papovaviruses, Robert Lehman for DNA polymerases and recombination, Charles McHenry for DNA polymerase III, David Korn for DNA polymerases $\alpha$ and $\beta$, Michael Chamberlin for RNA polymerases, Olke Uhlenbeck for ligases, Nicholas Cozzarelli and Joseph Coleman for gyrases and binding proteins, Stuart Linn for nucleases, Claude Paoletti for inhibitors, Gordon Lark for control of replication, George Stark for gene amplification, Robert Webster for phage M13, Bruce Alberts for phage T4, Charles Richardson for phage T7, David Clayton for mitochondria, Donald Helinski for plasmids, Jerard Hurwitz for adenoviruses, Michael Bishop and Inder Verma for retroviruses, Mehran Goulian for parvoviruses, Bernard Roizman for herpes viruses, William Robinson for hepadna viruses, Philip Hanawalt for DNA repair, Harrison Echols, Charles Radding, Robert Lehman, and Dale Kaiser for recombination, Leslie Orgel for prebiotic oligonucleotide synthesis, and Roberto Crea for chemical synthesis of DNA.

To all these colleagues, I and others who want to be au courant with DNA transactions are indebted. I take the responsibility, however, for editorial selections, revisions, and rather substantial additions to give the Supplement a uniform style. I am deeply grateful to Neil Patterson and John Staples of W. H. Freeman and Company for their eager acceptance and skillful execution of this experiment in publishing, to Betty Bray for typing this manuscript from crude handwritten copy, to Charlene Levering for artistic insights that enhanced the illustrations, to Patricia Brewer for excellent editorial styling, and to Roger Kornberg and Leroy Bertsch for careful review of the entire manuscript.

*January 1982*                                              *Arthur Kornberg*

1982 Supplement to
# DNA Replication

# S1

# Structure and Functions of DNA*

## Abstract

The past two years have been the most eventful in the history of DNA structure since the discovery of the double helix. X-ray crystallographic analysis, made possible only by the efficient new technologies of oligodeoxyribonucleotide synthesis, provides for the first time the structural details at individual base pairs along the helix. Cloning of long DNA molecules engineered to contain these synthetic sequences in defined arrangements further permits their natural behavior in solution to be analyzed. The x-ray crystal structures of DNA molecules so far determined have given several striking results: The essential features of A-DNA and B-DNA, as proposed and refined by fiber diffraction analysis, were confirmed;

---

*Section numbers from *DNA Replication* are listed below, with an S preceding those for which new information is provided in this Supplement.

novel properties of B-DNA, with regard to nucleotide conformation and flexibility of the double helix, have emerged; and a new family of DNA structures, termed Z-DNA, was discovered, which may hold clues to the control of gene expression.

## S1-1  DNA: Past and Present

Since *DNA Replication* appeared two years ago, the major currents of DNA research have continued and in several places have become much clearer. The ease of isolating, analyzing, synthesizing, and rearranging DNA sequences and genes and the ability to insert these recombinant DNAs into cells have fueled a stampede from all quarters of biology and has titillated chemists and industrialists. Interest in DNA transactions such as replication, repair, transposition, and viral multiplication has increased but remains predominantly focused on intact cellular systems. Too little attention has been given to identifying the proteins that translate the genetic message into cellular action, the agents for creating and controlling all genetic events. Still, as this Supplement will illustrate, progress in many areas has provided insights into the nature of DNA replication that impress all of us who observe the advancing front of DNA knowledge.

## S1-3  A Double Helical Structure[1-5]

Three families of DNA structure are now recognized, and examples of all three have been crystallized and subjected to x-ray structure determination (Fig. S1-1). The B family is the predominant form in solution. The crystal structure shows a classical, right-handed Watson–Crick double helix, with a striking degree of flexibility and sequence dependence of certain helix parameters. Conversion of the B structure to the A and Z forms usually attends lowering of water activity, for example, by addition of ethanol or at low humidity or high salt concentration, although some DNAs may adopt the Z structure in physiological conditions. A remarkable feature of Z-DNA is the left-handed sense of the double helix.

1. Viswamitra, M. A., Kennard, O., Shakked, Z., Jones, P. G., Sheldrick, G. M., Salisbury, S., and Falvello, L. (1978) *Nat. 273*, 687.
2. Wang, A. H.-J., Quigley, G. J., Kolpak, F. J., Crawford, J. L., van Boom, J. H., van der Marel, G., and Rich, A. (1979) *Nat. 282*, 680.
3. Drew, H., Takano, T., Tanaka, S., Itakura, K., and Dickerson, R. E. (1980) *Nat. 286*, 567.
4. Wing, R., Drew, H., Takano, T., Broka, C., Tanaka, S., Itakura, K., and Dickerson, R. E. (1980) *Nat. 287*, 755.
5. Dickerson, R. E., Drew, H. R., Conner, B. M., Wing, R. M., Fratini, A. V., and Kopka, M. L. (1982) *Science 216*, 475.

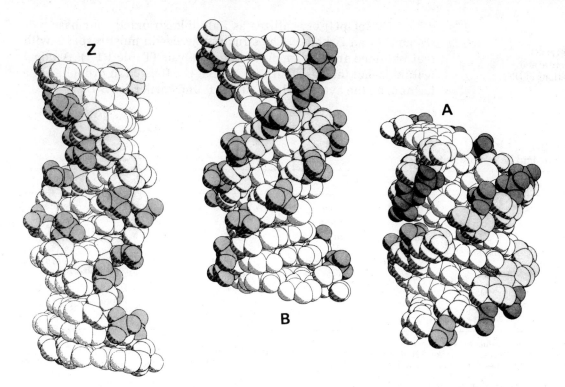

FIGURE S1-1
Space-filling drawings of A-, B-, and Z-DNA. Uniform-scale comparisons are made of equal lengths based on single-crystal structure analyses of A-DNA, three stacked CCGG tetramers; B-DNA, one CGCGAATTCGCG molecule; Z-DNA, three stacked CGCG tetramers. The tetramers have not been adjusted; they are stacked atop one another with the proper helical repeat, hence the phosphates needed to bridge between tetramers are absent. (Courtesy of Professor R. E. Dickerson.)

## A-DNA

The A structure is found in aggregates of DNA induced by 70–75 percent ethanol[6,7] and in fibers of DNA in a dehydrated state. The A structure is also the principal conformation of RNA under all conditions. (Polyribonucleotides, whether in an RNA–RNA duplex or a DNA–RNA heteroduplex, are unable to adopt the B conformation due to steric hindrance between the ribose 2′ hydroxyl group and the phosphate group of the adjacent nucleotide.) The tetranucleotide

6. Zimmerman, S. B., and Pheiffer, B. H. (1979) *JMB 135*, 1023.
7. Gray, D. M., Edmondson, S. P., Lang, D., and Vaughan, M. (1979) *N. A. Res.*, *6*, 2089.

d(iodo-CpCpGpG) crystallizes as a double-stranded four-base-pair segment of an A helix.[8] The structure agrees in most respects with that obtained from fiber diffraction analysis (Table S1-1). A novel feature is that the base pairs are not flat but twisted in propeller-like fashion, on the average about 18°, as found earlier for B-DNA.

TABLE S1-1
Nucleic acid helix parameters

| Family | A | | B | | Z |
|---|---|---|---|---|---|
| Environment | crystal | fiber | crystal | fiber | crystal |
| Helix sense | right | right | right | right | left |
| Glycosyl bonds | anti | anti | anti | anti | syn/anti |
| Sugar ring conformation (pucker) | C3'-endo | C3'-endo | variable | C2'-endo | alternating |
| Base pairs per turn | 10.7 | 11 | 9.7[a] | 10 | 12 |
| Rise per base pair (Å) | 2.3 | 2.6 | 3.3[a] | 3.4 | 3.7 |
| Family members | | A,A' | | B,C | Z,Z' |
| References | 1 | 2 | 3 | 2 | 4 |

[a]Average of eleven values for a 12-bp helix.

1. Conner, B. N., Takano, T., Tanaka, S., Itakura, K., and Dickerson, R. E. (1981) *Nat. 295*, 294.
2. Arnott, S. (1976) in *Organization and Expression of Chromosomes* (Allfrey, V. G., Bautz, E. F. K., McCarthy, B. J., Schimke, R. T., and Tissieres, A., eds.) Dahlem Conference, Berlin, p. 209.
3. Drew, H. R., Wing, R. M., Takano, T., Broka, C., Tanaka, S., Itakura, K., and Dickerson, R. E. (1981) *PNAS 78*, 2179.
4. Wang, A. H.-J., Quigley, G. J., Kolpak, F. J., Crawford, J. L. van Boom, J. H., van der Marel, G., and Rich, A. (1979) *Nat. 282*, 680.

B-DNA

A representative of the B family of structures, the principal conformation of DNA in solution, has been obtained in crystals of the dodecamer d(CpGpCpGpApApTpTpCpGpCpG).[9] This molecule is self-complementary and in duplex form contains a site for cleavage by the restriction endonuclease *Eco*RI. The structure of the duplex at 1.9 Å resolution[10] corresponds closely with the classical B helix deduced from fiber diffraction studies (Table S1-1). Minor differences from classical B helix geometry were expected: A propeller twist of the base pairs and smooth bending of the helix axis were proposed

8. Conner, B. N., Takano, T., Tanaka, S., Itakura, K., and Dickerson, R. E. (1981) *Nat. 295*, 294.
9. Wing, R., Drew, H., Takano, T., Broka, C., Tanaka, S., Itakura, K., and Dickerson, R. E. (1980) *Nat. 287*, 755; Dickerson, R. E., and Drew, H. R. (1981) *JMB 149*, 761; Drew, H. R., and Dickerson, R. E. (1981) *JMB 151.* 535; Dickerson, R. E., and Drew, H. R. (1981) *PNAS 78*, 7318.
10. Drew, H. R., Wing, R. M., Takano, T., Broka, C., Tanaka, S., Itakura, K., and Dickerson, R. E. (1981) *PNAS 78*, 2179.

on the basis of energy minimization calculations;[11] flexibility with regard both to bending of the helix axis and to the rotation from one base pair to the next around the helix axis was indicated by physico-chemical measurements on DNA in solution.[12] These expectations are borne out in the dodecamer structure.

The base pairs are twisted in propeller fashion from 5 to 19°, presumably because such twisting increases the overlap of a base with those above and below it in the same polynucleotide chain. The helix axis is bent, smoothly and without kinking, by a total of 19° over the full length of the dodecamer. This bending is probably a consequence of the packing arrangement in the crystal and not inherent in the dodecamer, but it illustrates the ease and mechanism of bending B-DNA. The rotation per base pair in the dodecamer ranges from 32 to 45°, with a mean of 37°, corresponding to 9.7 base pairs per turn of the helix (which may be compared with a value of 10.6 base pairs per turn estimated for all DNAs in solution except poly dA · poly dT).[13] In contrast with the bending of the helix axis, the nonuniform rotation per base pair seen in the crystal structure probably reflects the state of the dodecamer in solution, since the value of the rotation at a particular base pair step correlates in a striking way with the rate constant for cleavage at that point by DNase I.[14] It remains to be seen whether this sequence-dependent feature of B-DNA structure is recognized by other nucleases as well.

The flexibility and internal variation of the dodecamer structure can be traced to a variable conformation of the sugar-phosphate backbone, especially the deoxyribose ring (Fig. S1-2). A distribution extending from the C3'-endo ring conformation characteristic of the classical A helix to the C2'-endo conformation of the classical B helix is observed. An extreme case of this variable sugar conformation may be found in the "alternating-B" structure proposed for the alternating copolymer poly d(AT).[15] This proposal is based on the crystal structure of the tetranucleotide d(pApTpApT), which is organized in dinucleotide units, rather than as a single, 4-bp stretch of double helix.[16] The sugar conformation is C3'-endo within a unit and C2'-endo between units. While there is no direct evidence for the

11. Levitt, M. (1978) PNAS 75, 640.
12. Record, M. T., Jr., Mazur, S. J., Melançon, P., Roe, J.-H., Shaner, S. L., and Unger, L. (1981) ARB 50, 997.
13. Wang, J. C. (1979) PNAS 76, 200; Peck, L. J., and Wang, J. C. (1981) Nat. 292, 375; Rhodes, D., and Klug, A. (1981) Nat. 292, 378.
14. Lomonossoff, G. P., Butler, P. J. G., and Klug, A. (1981) JMB 149, 745.
15. Klug, A., Jack, A., Viswamitra, M. A., Kennard, O., Shakked, Z., and Steitz, T. A. (1979) JMB 131, 669.
16. Viswamitra, M. A., Kennard, O., Shakked, Z., Jones, P. G., Sheldrick, G. M., Salisbury, S., and Falvello, L. (1978) Nat. 273, 687.

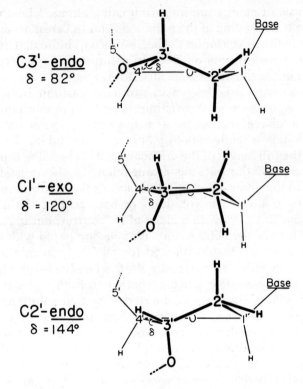

FIGURE S1-2
Relationship between deoxyribose sugar puckering and
the C5'-C4'-C3'-03' main chain torsion angle δ. This
torsion angle is a much more sensitive and easily
observed measure of sugar conformation than the
appearance of the 5-membered ring itself. (Courtesy of
Professor R. E. Dickerson.)

alternating-B structure of poly d(AT) copolymer in solution, such a
structure would explain why digestion of the copolymer to the limit
with DNase I results in a dinucleotide product.[17]

## Z-DNA

The proposed alternating-B structure foreshadowed the discovery of
an entirely new family of DNA structures in which sugar con-
formations alternate between C2'-endo and C3'-endo or C1'-exo.
The first members of this new family, termed Z-DNA because of
the zigzag course of the backbone, were found by x-ray crystal-

17. Scheffler, I. E., Elson, E. L., and Baldwin, R. L. (1968) *JMB 36*, 291.

lographic analysis of the hexanucleotide d(CpGpCpGpCpG)[18] and the tetranucleotide d(CpGpCpG).[19] The structures of the duplex forms of these molecules are, like that of d(pApTpApT), based on a dinucleotide unit, but they differ from A-, B-, and alternating-B DNA in two fundamental respects. First, the helix sense is left- rather than right-handed. Second, the glycosyl bonds alternate between *syn* and *anti* orientations, in contrast with the all-*anti* conformation of previously studied DNAs.

To what extent might the Z structure occur in nature in DNA sequences of interest? It can be argued that the Z structure is limited to alternating purine–pyrimidine sequences, because steric repulsion prevents a pyrimidine from adopting the *syn* conformation.[20] Moreover, a Z-compatible sequence may be forced into the B structure by a neighboring region of B-DNA (as occurs in the dodecamer discussed above, which is found in the B structure in spite of the tendency of eight of twelve residues to form a Z helix on their own in the same conditions). Factors that might favor the Z structure include methylation[21,22] and specific DNA-binding proteins. The search for Z-DNA in a physiologically important context should be facilitated by the isolation of a remarkably specific antibody that recognizes only the Z form of DNA.[23] The antibodies bind specifically and regularly to interband regions of polytene chromosomes of *Drosophila*.[24] Because cloned segments of poly d(GC) in plasmids undergo a reversible B to Z transformation,[25] it seems possible that similar changes in conformation in chromosomes may accompany local environmental influences and have far-reaching effects.

## S1-6  Shape

The DNA helix behaves as a stiff rod only for short segments in the size range of the persistence length (about 500 Å, or 150 base pairs).[26] Even such short DNA molecules show a finite probability of bending: Restriction fragments as short as 242 base pairs have been closed

18. Wang, A. H.-J., Quigley, G. J., Kolpak, F. J., Crawford, J. L., van Boom, J. H., van der Marel, G., and Rich, A. (1979) *Nat. 282,* 680.
19. Drew, H., Takano, T., Tanaka, S., Itakura, K., and Dickerson, R. E. (1980) *Nat. 286,* 567; Drew, H. R., and Dickerson, R. E. (1981) *JMB 152,* 723.
20. Drew, H., Takano, T., Tanaka, S., Itakura, K., and Dickerson, R. E. (1980) *Nat. 286, 567.*
21. Behe, M., and Felsenfeld, G. (1981) *PNAS 78,* 1619.
22. Möller, A., Nordheim, A., Nichols, S. R., and Rich, A. (1981) *PNAS 78,* 4777.
23. Lafer, E. M., Möller, A., Nordheim, A., Stollar, B. D., and Rich, A. (1981) *PNAS 78,* 3546.
24. Nordheim, A., Pardue, M. L., Lafer, E. M., Möller, A., Stollar, B. D., and Rich, A. (1981) *Nat. 294,* 417.
25. Klysik, J., Stirdivant, S. M., Larson, J. E., Hart, P. A., and Wells, R. D. (1981) *Nat. 290,* 672; Peck, L., Nordheim, A., Rich, A., and Wang, J. C., personal communication.
26. Hagerman, P. J. (1981) *Biopolymers 20,* 1503.

covalently into circles by T4 DNA ligase.[27] This flexibility of the double helix is also illustrated by the folding of large DNA molecules into compact structures. The tight winding of DNA about a histone octamer to form the nucleosome (Section S9-7) is effected by the binding energies of the strong ionic interactions. Also at very low DNA concentrations, monomolecular condensation can be induced by polyamines[28] or even by the trivalent ion $Co^{3+}(NH_3)_6$.[29] The tightly wound condensates have diameters close to those of the phage head interior.

Evidence for a sequence-specific fold in DNA, the so-called cruciform or double hairpin structure, has recently been found.[30] The supercoiled forms of ColE1, pBR322, and $\phi$X174 RF DNA all contain one major site of cleavage by the single-strand-specific nuclease S1. In every case this site lies in an inverted repeat region, suggestive of the double hairpin loops previously proposed to accommodate the strain of negative supercoiling with the minimum loss of base pairing. Remarkably, the site of cleavage in $\phi$X174 DNA is the same as the site of binding of a prepriming replication protein to a single strand of DNA and supports the idea that this protein recognizes a hairpin-loop structure.[31] Hairpin-loop formation by an inverted repeat is a local property that can be transmitted from one supercoiled DNA to another by moving the responsible region through recombinant DNA manipulation.[32]

27. Shore, D., Langowski, J., and Baldwin, R. L. (1981) *PNAS 78,* 4833.
28. Gosule, L. C., and Schellman, J. A. (1976) *Nat. 259,* 333.
29. Widom, J., and Baldwin, R. L. (1980) *JMB 144,* 431.
30. Lilley, D. M. J. (1980) *PNAS 77,* 6468.
31. Shlomai, J., and Kornberg, A. (1980) *PNAS 77,* 799.
32. Lilley, D. M. J. (1981) *N. A. Res. 9,* 1271.

# S2

# Biosynthesis of DNA Precursors

## Abstract

Purine and pyrimidine biosynthetic pathways and the one-carbon metabolism that feeds them are served by multifunctional enzymes and multienzyme complexes that efficiently channel and preserve labile intermediates. New findings include the following: (i) More refined genetic analysis reveals complexities in the regulatory control of the pathways. (ii) Phage-induced replication systems have assemblies that guide even remote DNA precursors directly to the site of polymerization. (iii) Ribonucleotide reduction, a crucial regulatory stage in DNA synthesis, is catalyzed and controlled in mammals much as in *E. coli*. The reductase mechanism, especially the action of the tyrosyl radical, is under active study. (iv) Inhibition of thymidylate synthesis by methotrexate elevates dUTP to levels that result in harmful incorporation of uracil in DNA. (v) Failure to remove purine nucleotide breakdown products may lead to abnormal salvage, as happens when reduced levels of adenosine deaminase or purine nucleoside phosphorylase produce severe immunodeficiency diseases. However, retention of uric acid, the ultimate purine degradation product in humans, long reviled as the cause of gouty arthritis and renal disease, may prove a friend as a potent antioxidant in combating mutagenic agents that affect aging and cause cancer.

## S2-2    Purine Nucleotide Synthesis de Novo

### PRPP Synthetase

PRPP synthetase, an enzyme from human erythrocytes, is regulated by the state of its aggregation. In vitro, the enzyme varies in size from a dimer to aggregates of 16 or 32 subunits.[1] Magnesium ATP and other purine nucleotides promote aggregation, while 2,3-bisphosphoglycerate, an inhibitor, promotes dissociation; the smaller forms have only 4 percent of the catalytic activity of the very large.

### Genes and Regulation of Purine Biosynthesis

Genes are known in S. typhimurium and E. coli for all but one of the twelve enzymes that convert PRPP to GMP and AMP (Table S2-1).[2] Most of these genes are physically unlinked, although several respond to a common repressor, isolated on the basis of binding to specific operator regions.[3] ATP and GTP may each act as corepressors with differential effects on reactions before the IMP branchpoint. What once appeared to be a simple and decisive regulation at IMP will likely prove more complex both at this level and at earlier points in the de novo pathway.[4] In yeast, relatively little is known about the

1. Becker, M. A., Meyer, L. J., Huisman, W. H., Lazar, C., and Adams, W. B. (1977) JBC 252, 3911; Meyer, L. J., and Becker, M. A. (1977) JBC 252, 3919.
2. Bachmann, B. J., and Low, K. B. (1980) Microbiol. Rev. 44, 1; Sanderson, K. E., and Hartman, P. E. (1978) Microbiol. Rev. 42, 471.
3. Gots, J. S., Benson, C. E., Jochimsen, B., and Koduri, K. R. (1976) in Microbial Models and Regulatory Elements in the Control of Purine Metabolism, pp. 23–40, Ciba Foundation Symposium, Purine and Pyrimidine Metabolism; Levine, R. A., and Taylor, M. W. (1982) J. Bact. 149, 923; Koduri, K. R., and Gots, J. S. (1980) JBC 255, 9594.
4. Levine, R. A., and Taylor, M. W. (1981) MGG 181, 313: (1982) J. Bact. 149, 923.

TABLE S2-1
Genes of purine biosynthesis in *E. coli*

| Enzyme | Reaction | Gene | Map position[a] |
|---|---|---|---|
| Amidophosphoribosyl-transferase | 5-phosphoribosyl-α-pyrophosphate (PRPP) ↓ 5-phosphoribosylamine | *purF* | 49 |
| GAR synthetase | ↓ glycinamide ribotide (GAR) | *purD* | 89 |
| GAR transformylase | ↓ formylglycinamide ribotide (FGAR) | ? | |
| FGAM synthetase | ↓ formylglycinamidine ribotide (FGAM) | *purG* | 53 |
| AIR synthetase | ↓ 5-aminoimidazole ribotide (AIR) | *purI* | 55 |
| AIR carboxylase | ↓ carboxyaminoimidazole ribotide (CAIR) | *purE* | 12 |
| SACAIR synthetase | ↓ 5-aminoimidazole-4-(N-succinylo-carboxamide) ribotide (SACAIR) | *purC* | 53 |
| Adenylosuccinate lyase | ↓ 5-aminoimidazole-4-carboxamide ribotide (AICAR) | *purB* | 25 |
| AICAR transformylase | ↓ 5-formaminoimidazole-4-carboxamide ribotide (FAICAR) ↓ IMP | *purH* | 89 |
| Adenylosuccinate synthetase | IMP ↓ Adenylosuccinate | *purA* | 94 |
| Adenylosuccinate lyase | ↓ AMP | *purB* | 25 |
| IMP dehydrogenase | IMP ↓ XMP | *guaB* | 53 |
| GMP synthetase (xanthylate aminase) | ↓ GMP | *guaA* | 53 |

[a]From Bachman, B. J., and Low, K. B. (1980) *Microbiol. Rev.* 44, 1.

enzymes or their regulation, even though at least seven of the purine biosynthetic genes have long been identified.[5]

## Glutamine-PRPP Amidotransferase

An iron-sulfur cluster is found at the active center of glutamine-PRPP amidotransferase, the first enzyme of purine biosynthesis of *B. subtilis*.[6] This enzyme is the target of complex regulatory controls, one of which is inactivation of the enzyme by molecular oxygen when cells enter stationary phase.

## One-Carbon Folate Cofactor

The true substrate for glycinamide ribonucleotide (GAR) transformylase is L(-)-10-formyl-$H_4$folate, and not 5,10-methenyl-$H_4$folate;[7] the latter functions in standard enzyme assays because GAR transformylase is active only when in a complex with the trifunctional folate enzyme that can convert 5,10-methenyl-$H_4$folate to 10-formyl-$H_4$folate (Section S2-14).

## Transformylase-Trifunctional Folate Enzyme Complex

Both GAR transformylase and aminoimidazolecarboxamide ribonucleotide (AICAR) transformylase from chicken liver have been isolated as a complex with the trifunctional folate enzyme that synthesizes the folate cofactors (Figure S2-1, below).[8] This proximity of the different enzyme catalytic centers provides the ready supply of a labile cofactor from the folate synthesizing enzyme to the GAR or AICAR transformylase, and in addition the integrity of this multienzyme complex is necessary for GAR transformylase to have activity[9] (Section S2-14). Not only is the generation of formylfolate coupled to its transfer, but beyond that, the formyl transferring activity (AICAR transformylase) in chicken liver is linked in a single polypeptide to inosinicase activity[10] to ensure the next and final step in purine biosynthesis.

5. Armitt, S., and Wood, R. A. (1970) *Genet. Res. Camb. 15,* 7.
6. Wong, J. Y., Bernlohr, D. A., Turnbough, C. L., and Switzer, R. L. (1981) *B. 20,* 5669; Bernlohr, D. A., and Switzer, R. L. (1981) *B. 20,* 5675.
7. Dev, I. K., and Harvey, R. J. (1978) *JBC 253,* 4242; Smith, G. K., Benkovic, P. A., and Benkovic, S. J. (1981) *B. 20,* 4034.
8. Caperelli, C. A., Benkovic, P. A., Chettur, G., and Benkovic, S. J. (1980) *JBC 255,* 1885.
9. Smith, G. K., Mueller, W. T., Wasserman, G. F., Taylor, W. D., and Benkovic, S. J. (1980) *B. 19,* 4313.
10. Mueller. W. T., and Benkovic, S. J. (1981) *B. 20,* 337.

## The Trifunctional Enzyme

The protein that initiates eukaryotic pyrimidine biosynthesis can be cleaved into smaller peptides to yield peptide fragments each with a single enzyme activity.[12] The transcarbamylase domain appears to be essential for the association of the 210-kdal polypeptide chains to yield trimers, hexamers, etc., of the native protein.[13] The domains of the aggregated protein are arranged such that the two intermediates, carbamyl phosphate and carbamyl aspartate, are normally not released into the cytosol but are preferentially converted to the final product, dihydroorotate, which then is rapidly converted to UMP by the last three enzymes of this pathway.[14] Carbamyl aspartate is not protonated[15] on the amide N (as shown in *DNA Replication*, Fig. 2-7, p. 49), but a proton participates in the dehydration to dihydroorotate, as does zinc.[16]

## The Bifunctional Enzyme

UMP synthetase, containing orotate phosphoribosyltransferase and orotidylate decarboxylase, has been purified to homogeneity as a polypeptide of 51 kdal.[17] It can exist as a monomer with phosphoribosyltransferase activity but no decarboxylase activity. However, orotidine monophosphate (OMP), the product of the phosphoribosyltransferase, converts the monomer to a conformationally active dimer that has decarboxylase activity. The barbituric acid analog of OMP binds the yeast decarboxylase with extraordinary affinity, $10^5$ times as strongly as OMP, suggesting a possible structure for the transition state.[18]

11. Jones, M. E. (1980) *ARB 49*, 253.
12. Davidson, J. N., Rumsby, P. C., and Tamaren, J. (1981) *JBC 256*, 5220; Mally, M. I., Grayson, D. R., and Evans, D. R. (1981) *PNAS 78*, 6647.
13. Dev, I. K., and Harvey, R. J. (1978) *JBC 253*, 4242; Smith, G. K., Benkovic, P. A., and Benkovic, S. J. (1981) *B. 20*, 4034; Coleman, P. F., Suttle, D. P., and Stark, G. R. (1977) *JBC 252*, 6379.
14. Jones, M. E. (1980) *ARB 49*, 253; Christopherson, R. I., and Jones, M. E. (1980) *JBC 255*, 11381; Mally, M. I., Grayson, D. R., and Evans, D. R. (1980) *JBC 255*, 11372.
15. Christopherson, R. I., and Jones, M. E. (1979) *JBC 254*, 12506.
16. Christopherson, R. I., and Jones, M. E. (1980) *JBC 255*, 3358; Taylor, W. H., Taylor, M. L., Balch, W. E., and Gilchrist, P. S. (1976) *J. Bact. 127*, 863.
17. McClard, R. W., Black, M. J., Livingstone, L. R., and Jones, M. E. (1980) *B. 19*, 4699; Traut, T. W., Payne, R. C., and Jones, M. E. (1980) *B. 19*, 6062; Traut, T. W., and Payne, R. C. (1980) *B. 19*, 6068.
18. Levine, H. L., Brody, R. S., and Westheimer, F. H. (1980) *B. 19*, 4993.

When the overall pathway of pyrimidine biosynthesis functions in intact cells, the five intermediates—carbamyl phosphate, carbamyl aspartate, dihydroorotate, orotate, and orotidylate—are held at micromolar levels in part because of channeling by the trifunctional polypeptide that initiates pyrimidine synthesis and the bifunctional polypeptide, UMP synthetase, that completes the de novo formation of UMP.[19] Channeling of OMP by UMP synthetase may be physiologically significant since OMP can be degraded to orotidine by a pyrimidine nucleotidase or possibly by a specific OMP nucleotidase.[20] Orotidine can be converted back to orotate.[21] Inhibition of UMP synthetase leads to a large accumulation of carbamyl aspartate in cultured cells with a lesser accumulation of orotidine and orotate; patients with orotic aciduria also excrete carbamyl aspartate.[22]

## S2-6    Ribonucleotide Reduction to Deoxyribonucleotide

### Mammalian Reductase

Ribonucleotide reductase from calf thymus has been extensively purified and characterized.[23] The enzyme consists of two nonidentical subunits, M1 and M2 (M for mammalian), which in most respects appear to be the correlates of the B1 and B2 subunits of *E. coli*. Regulation via the M1 subunit shows the same characteristic features as does the bacterial enzyme, but shows them even more sharply. The two classes of allosteric sites in M1 were defined by binding experiments and by mutations affecting either class of sites in a cultured mouse cell line.[24] These mutations increase the overall mutation rates of cells[25] by affecting the dNTP levels; the reductase

19. Christopherson, R. I., Traut, T. W., and Jones, M. E. (1981) *Curr. Topics Cell Regul. 18,* 59.
20. Traut, T. W. (1980) *Arch. Biochem. Biophys. 200,* 590; El Kouni, M. H., and Cha, S. (1981) *Fed. Proc. 40,* 924.
21. Christopherson, R. I., Traut, T. W., and Jones, M. E. (1981) *Curr. Topics Cell Regul. 18,* 59.
22. Jones, M. E. (1980) *ARB 49,* 253; Christopherson, R. I., and Jones, M. E. (1980) *JBC 255,* 11381; Mally, M. I., Grayson, D. R., and Evans, D. R. (1980) *JBC 255,* 11372.
23. Thelander, L., and Reichard, P. (1979) *ARB 48,* 133; Thelander, L., Eriksson, S., and Åkerman, M. (1980) *JBC 255,* 7426.
24. Eriksson, S., Thelander, L., and Åkerman, M. (1979) *B. 18,* 2948; Ullman, B., Clift, S. M., Gudas, L. J., Levinson, B. B., Wormsted, M. A., and Martin, D. W., Jr. (1980) *JBC 255,* 8308; Eriksson, S., Gudas, L. J., Clift, S. M., Caras, I. W., Ullman, B., and Martin, D. W., Jr. (1981) *JBC 256,* 10193.
25. Weinberg, G., Ullman, B., and Martin, D. W., Jr. (1981) *PNAS 78,* 2447.

gene may thus function as a mutator.[26] Variation in reductase levels in response to the cell cycle depends on the M2 activity level.[27]

## Mechanism

The presence of a free tyrosyl radical in the M2 subunit was demonstrated by EPR spectroscopy of mouse fibroblast cells overproducing the radical during the course of developing resistance against hydroxyurea.[28] Regulating the degree of radical formation ("radicalization" of the reductase) may offer an additional mechanism for controlling both the mammalian and *E. coli* enzymes.

A mechanism[29] for the *E. coli* reductase involving formation of a very short-lived substrate radical proposes abstraction of a hydrogen atom from the 3'-carbon, possibly by the tyrosyl radical, and its restoration following the loss of the 2'-hydroxyl. In favor of this scheme is the observation of an interaction between the reductase and the substrate analog, 2'-azido CDP, resulting in the transient appearance of an electron paramagnetic resonance signal localized to the nucleotide.[30] Presumably the lifetime of the radical is extended by the azido substituent. A similar abstraction of hydrogen from the 3' position appears to occur with the *Lactobacillus* reductase, but no evidence for participation of the $B_{12}$ cofactor has been obtained.[31]

## S2-8   dUTPase in Thymidylate Biosynthesis

Under normal circumstances dUTPase operates very efficiently, leaving no measurable dUTP pools in animal cells.[32] However, dUTP was demonstrated after infection of 3T6 cells with polyoma virus[33] or by inhibiting thymidylate synthesis in lymphoblast cells with methotrexate.[34] After polyoma infection, the dUTP pool amounted to no more than 0.4 percent of the dTTP pool and did not increase after hydroxyurea treatment, and so could not have contributed

26. Fersht, A. R., and Knill-Jones, J. W. (1981) *PNAS 78*, 4251.
27. Eriksson, S., and Martin, D. W., Jr. (1981) *JBC 256*, 9436.
28. Åkerblom, L., Ehrenberg, A., Gräslund, A., Lankinen, H., Reichard, P., and Thelander, L. (1981) *PNAS 78*, 2159.
29. Stubbe, J. A., and Ackles, D. (1980) *JBC 255*, 8027.
30. Sjöberg, B. M., personal communication.
31. Stubbe, J. A., Ackles, D., Segal, R., and Blakley, R. L. (1981) *JBC 256*, 4843.
32. Nilsson, S., Reichard, P., and Skoog, L. (1980) *JBC 255*, 9552; Goulian, M., Bleile, B., and Tseng, B. Y. (1980) *JBC 255*, 10630.
33. Nilsson, S., Reichard, P., and Skoog, L. (1980) *JBC 255*, 9552.
34. Goulian, M., Bleile, B., and Tseng, B. Y. (1980) *PNAS 77*, 1956; *JBC 255*, 10630.

significantly to the generation of short (4S) Okazaki fragments in mammalian cells. After methotrexate, however, dUTP levels reach 20 percent of that of dTTP, and uracil incorporation and excision might well contribute to the deranging effects of this drug.

## S2-9   Salvage Pathways of Nucleotide Synthesis

Appreciation of the vital importance of various salvage pathways has been enlarged by more discoveries: (i) diseases due to enzyme deficiencies, (ii) toxic effects of precursors and analogs, and (iii) instances in which deficiencies and toxic effects are combined. These discoveries illustrate another dimension of salvage pathways, namely, their homeostatic function. They participate in a complex interplay of de novo, degradative, salvage, and excretory pathways to cope with changing environmental stresses, nutritional factors, and altered enzymic capacities.

### Human Immunodeficiency Diseases

Purine salvage and disposal enzymes, adenosine deaminase and purine nucleoside phosphorylase, have special significance for proper immune function in lymphocyte differentiation.[35] Adenosine deaminase deficiency leads to a severe combined immunodeficiency disease resulting from a defect in both T-cell and B-cell development. A deficiency of purine nucleoside phosphorylase does not affect B-cell development but results in impaired T-cell development and function.

_Adenosine deaminase deficiency_ results in a 100-fold elevation of the dATP level in T-cells, but not in B-cells. The elevated dATP, resulting from the accumulation of deoxyadenosine, is a powerful negative regulator of ribonucleotide reductase and depresses the level of the other deoxynucleoside triphosphates (dNTPs). The basis for B-cell toxicity is less clear. Increased adenosine levels may inhibit S-adenosylhomocysteine hydrolase (Fig. S12-3). The resulting increased levels of S-adenosylhomocysteine would inhibit a number of essential methylation reactions. The reason that dNTPs do not accumulate in B-cells may in part be a higher level of deoxynucleotidase activity.[36]

---

35. Barton, R. W., and Goldschneider, I. (1979) _Molec. Cell. Biochem. 28_, 135; Hershfield, M. S., and Kredich, N. M. (1978) _Science 202_, 757; Martin, D. W., Jr., and Gelfand, E. W. (1981) _ARB 50_, 845.
36. Carson, D. A., Kaye, J., and Wasson, D. B. (1981) _J. Immunol. 126_, 348.

_Purine nucleoside phosphorylase deficiency_ results in a huge accumulation of dGTP in T-cells. A suggested mechanism[37] is based on the observed differential metabolism of deoxyguanosine in B-cells and T-cells and on the toxicity of deoxyguanosine only for T-lymphoblastoid cells. The major route of deoxyguanosine metabolism in B-cells is degradation to guanine and then conversion to GMP, whereas in T-cells it is converted directly by a nonspecific deoxycytidine kinase to dGMP and then phosphorylated to dGTP to become a negative effector for ribonucleotide reductase. Supporting this mechanism is the toxicity of 8-aminoguanosine, a purine nucleoside phosphorylase inhibitor.[38] T-cells but not B-cells are affected with a fourfold increase in dGTP even at low levels of deoxyguanosine.

## Other Purine Salvage Enzymes in Humans

The important purine salvage enzyme hypoxanthine-guanine phosphoribosyltransferase (HG-PRTase) from human erythrocytes and fibroblasts and its mutant forms have now been characterized.[39] Still another purine salvage enzyme is significant in human malignant tumor cell lines that lack the phosphorylase[40] that salvages adenine from methylthioadenosine. The latter is a byproduct in the conversion of S-adenosylmethionine in polyamine synthesis (Fig. S12-3). This deficiency makes such tumor cells vulnerable to certain chemotherapeutic approaches.

## Adenine Toxicity in _E. coli_

Growth inhibition by very low levels of adenine is observed in mutants defective in hypoxanthine and guanine phosphoribosyl transferases and is relieved by secondary mutations that remove either the adenine phosphoribosyl transferase or a repressor (_purR_) that acts at several loci before the IMP branchpoint.[41] The toxic effects have been attributed to a depletion of guanine nucleotide pools and suggest that the salvage of guanine to form GMP has a major role in maintaining GTP levels in the face of circumstances such as adenine excess.

37. Osborne, W. R. A. (1981) _TIBS 6_, 80.
38. Kazmers, I. S., Mitchell, B. S., Dadonna, P. E., Wotring, L. L., Townsend, L. B., and Kelley, W. N. (1981) _Science 214_, 1137.
39. Holden, J. A., and Kelley, W. N. (1978) _JBC 253_, 4459; Johnson, G. G., Eisenberg, L. R., and Migeon, B. R. (1979) _Science 203_, 174; Wilson, J. M., Baugher, B. W., Landa, L., and Kelley, W. N. (1981) _JBC 256_, 10306.
40. Kamatani, N., Nelson-Rees, W. A., and Carson, D. A. (1981) _PNAS 78_, 1219.
41. Levine, R. A., and Taylor, M. W. (1981) _MGG 181_, 313; (1982) _J. Bact. 149_, 923.

The de novo purine biosynthetic pathway in Chinese hamster cells is about five times more sensitive to inhibition by exogenous adenine than is the pathway in E. coli.[42] Adenine becomes lethal for these cells when coformycin, an adenosine analog, is also present in the medium, an effect that is reversible by hypoxanthine. These results suggest that the cells are starved of vitally needed IMP by simultaneous shutoff of the purine de novo biosynthesis pathway and the deamination of adenosine and adenylate.[43] Still another source of adenine toxicity is the depression of pyrimidine as well as purine nucleotide synthesis that results from lowering the PRPP level.[44]

## S2-13 Virus-Induced Patterns of Nucleotide Biosynthesis

T4 phage have evolved a highly efficient assembly line containing seven enzymes for synthesizing deoxyribonucleoside triphosphates from more distant precursors.[45] This multienzyme complex contains six phage enzymes (dCMP deaminase, dCMP hydroxymethylase, ribonucleotide reductase, dTMP synthetase, dNMP kinase, and dUTPase) and the E. coli nucleoside diphosphate kinase. By channeling precursors, dNMPs are incorporated into DNA more rapidly than exogenous dNTPs, and even rNDPs are incorporated more rapidly than dNTPs. This channeling may be advantageous to ensure maximum incorporation of hydroxymethyl dCMP specifically needed for T4 phage DNA synthesis (see Fig. 2-24). Dihydrofolate reductase[46] is another phage-encoded enzyme that may interact with replication proteins and be regulated along with them; it is also a structural element of the phage tail baseplate.

## S2-14 One-Carbon Metabolism

The one-carbon donor in purine biosynthesis is exclusively 10-formyl-$H_4$folate (Section S2-2), contrary to the scheme in Figure 2-27. The naturally occurring polyglutamate forms of folate are the pre-

42. Taylor, M. W., Olivelle, S., Levine, R. A., Coy, K., Hershey, H., Gupta, K. C., and Zawistowich, L. (1982) JBC 257, 377.
43. Debatisse, M., Berry, M., and Buttin, G. (1981) J. Cell. Physiol. 106, 1.
44. Chen, J.-J., and Jones, M. E. (1979) JBC 254, 2697.
45. Mathews, C. K., North, T. W., and Reddy, G. P. V. (1979) Adv. Enz. Regul. 17, 133; Allen, J. R., Reddy, G. P. V., Lasser, G. W., and Mathews, C. K. (1980) JBC 255, 7583.
46. Purohit, S., Bestwick, R. K., Lasser, G. W., Rogers, C. M., and Mathews, C. K. (1981) JBC 256, 9121.

ferred cofactors for many enzymes.[47] The pentaglutamyl form of 5,10-methylene-$H_4$folate has a $K_m$ of 7 $\mu$M for thymidylate synthetase of yeast compared to a value of 70 $\mu$M for the monoglutamyl form.[48]

Three catalytic activities in the processing of 5,10-methylene-$H_4$folate to $H_4$folate are on a single trifunctional protein[49] (enzymes 1, 2, and 3 in Figures S2-1 and S2-2); two of these catalytic centers are able to channel 5,10-methylene-$H_4$folate to 10-formyl-$H_4$folate.[50] This trifunctional protein exists in a multienzyme aggregate that also contains serine hydroxymethyl transferase (enzyme 4) which, from $H_4$folate, regenerates the 5,10-methylene-$H_4$folate required for dTMP synthesis.[51] The complex also contains the two transformylase enzymes (enzymes 5 and 6) of de novo purine biosynthesis. Some of the folate cofactors are quite labile in an aqueous solvent, and the multienzyme complex with its ability to channel the $H_4$folate intermediate is an elegant solution to this problem. A similar purpose may be served by linkage of the dihydrofolate reductase and thymidylate

47. Brody, T., Watson, J. E., and Stokstad, E. L. R. (1982) *B.* 21, 276.
48. Bisson, L. F., and Thorner, J. (1981) *JBC* 256, 12456.
49. Smith, G. K., Mueller, W. T., Wasserman, G. F., Taylor, W. D., and Benkovic, S. J. (1980) *B.* 19, 4313; Paukert, J. L., Straus, L. D., and Rabinowitz, J. C. (1976) *JBC* 251, 5104; Tan, L. U. L., Drury, E. J., and MacKenzie, R. E. (1977) *JBC* 252, 1117.
50. Cohen, L., and MacKenzie, R. E. (1978) *BBA* 522, 311.
51. Caperelli, C. A., Benkovic, P. A., Chettur, G., and Benkovic, S. J. (1980) *JBC* 255, 1885.

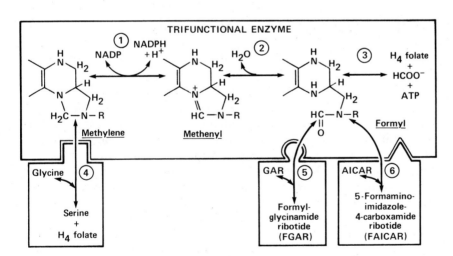

FIGURE S2-1
Trifunctional enzyme interconverting methylene, methenyl, and formyl tetrahydrofolate coenzymes and complexing with transferases to glycine, glycinamide ribotide (GAR), and 5-aminoimidazole-4-carboxamide ribotide (AICAR). (Courtesy of Professor M. E. Jones.)

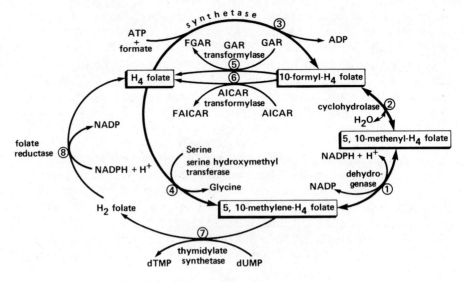

FIGURE S2-2
Tetrahydrofolate coenzyme cycle linked to transferases, thymidylate synthetase, and
dihydrofolate (folate) reductase. (Courtesy of Professor M. E. Jones.)

synthetase activities of the protozoan *Crithidia fasciculata* in a
single polypeptide.[52]

## S2-15   Traffic Patterns on the Pathways

The pathways of nucleotide metabolism can be likened to a city's
traffic patterns. Salvage routes can be used instead of de novo
pathways either as energetically attractive alternatives or out of
necessity. Problems arise when traffic pours in (e.g., excess adenine
or thymidine), some routes are blocked and the capacities of available pathways are exceeded. Pileups may even be aggravated by
absence or intervention of repressor and allosteric controls, and
the system then fails at several crucial places.

Eukaryotic cells have facilitated pathways of nucleotide metabolism by two strategies (Sections S2-2, S2-3, S2-13, S2-14). Enzymes
with sequential functions that are discrete polypeptides in prokaryotes
are often connected in a multifunctional polypeptide in eukaryotes.
Efficient channeling and preservation of labile intermediates is also
achieved by multienzyme complexes, one containing as many as

52. Ferone, R., and Roland, S. (1980) *PNAS 77*, 5802.

seven distinct enzymes. Finally, there is at least one example of a multienzyme complex that in turn contains a multifunctional protein (Section S2-14).

The human trait of maintaining high blood uric acid levels that may lead to gout has been regarded as an evolutionary misfortune (see p. 85) but may instead be an advantage should the potent antioxidant property of uric acid prove to be an important defense against oxidant- and radical-caused aging and cancer.[53]

The significance of degradative and excretory mechanisms as metabolic controls is now appreciated in the disposal of nucleosides (Section S2-9) and excess one-carbon units.[54]

53. Ames, B. N., Cathcart, R., Schwiers, E., and Hochstein, P. (1981) *PNAS 78*, 6858.
54. Scrutton, M. C., and Beis, I. (1979) *BJ 177*, 833.

# S3

# DNA Synthesis

# S4

# DNA Polymerase I of *E. coli*

# Abstract

The basic features of polymerase action are still well represented by the prototypical DNA polymerase I of E. coli. Routine assays of polymerase activity have become more facile, miniaturized, and specialized to assess processivity. Cellular location of a polymerase can be defined by immunocytochemistry (Section S6-1). The nucleotide sequence of the cloned gene that encodes DNA polymerase I provides the definitive amino acid sequence and clues to secondary structure, but, for lack of adequate crystallographic information, the three-dimensional shape of the enzyme and its interactions with nucleotides and DNA remain obscure. In addition to its essential roles in DNA repair and replication, the enzyme is required for transposition, the movement of transposable elements to new locations in a genome.

## S4-1  Isolation and Physicochemical Properties[1,2]

The amino acid sequence of DNA polymerase I has been deduced from the nucleotide sequence of the cloned polA gene and verified by limited amino acid sequence analysis of the large fragment (p. 139).[3] The sequence defines a polypeptide of 911 residues (102 kdal), consistent with earlier chemical analyses except that two cysteine residues are present, rather than three, one in the large fragment and the other in the small. The proteolytic cleavage that yields the two fragments (p. 139) is between a threonine at residue position 323 and valine at 324. This point of cleavage is close to the site of the polA1 mutation (Fig. 4-36) in which codon 342 is changed from trp (TGG) to amber (TAG). The finding of a polypeptide in extracts of the polA1 mutant having 5′→3′ exonuclease activity and the size of the small fragment are consistent with this assignment of the site of the amber mutation.[4] It seems even clearer now that the polA1 amber fragment is the small fragment and once formed is insusceptible to proteolysis. Predictions of secondary and tertiary structures based on amino acid sequence[5] do not disclose a hinged region in the vicinity of the proteolytic cleavage site.

1. Kornberg, A. (1981) Enzymes 14, 3.
2. Lehman, I. R. (1981) Enzymes 14, 15.
3. Joyce, C. M., Kelley, W. S., and Grindley, N. D. F. (1982) JBC 257, 1958.
4. Lehman, I. R., and Chien, J. R. (1973) JBC 248, 7717.
5. Brown, W. E., Stump, K. H., and Kelley, W. S. (1982) JBC 257, 1965.

S4-8   Polymerization Cycle and Processivity

## Ordered Mechanism

The polymerase reaction appears to follow an ordered mechanism as described for eukaryotic DNA polymerases (Sections S6-2, S6-3). Enzyme and template-primer combine first to form a binary complex, which subsequently binds and reacts with a nucleoside triphosphate. In pulse-chase experiments,[6] DNA polymerase I was incubated with poly dA · oligo(dT)$_{10}$; then Mg · dTTP and an excess of calf thymus DNA were added. The complex of enzyme and template-primer reacted with dTTP to elongate the primer for one processive cycle before the enzyme dissociated and became bound to the calf thymus DNA. Similar experiments in which the enzyme was incubated with $^3$H-dTTP before the template-primer and excess unlabeled dTTP were added led to no incorporation of $^3$H-dTTP, showing the lack of initial complex formation between enzyme and nucleoside triphosphate.

## Processivity

New methods have been introduced to measure processivity. An enzymatic analysis determines the value directly by measuring the number of chain ends elongated and the extent of elongation.[7] Processivity values can be determined even when two classes of product are made[8] and when synthesis is extremely processive.[9] The processivity of a polymerase is related to its physiological role. In replication, repair, recombination, and transposition, some polymerases have the responsibility for filling gaps while others synthesize long strands, and the frequency with which the polymerases dissociate can be correlated with these functions (Table S4-1). For example, the patch lengths observed in the repair of ultraviolet lesions in E. coli by the various DNA polymerases are proportional to their processivity.[10]

6. Bryant, R., and Benkovic, S., personal communication.
7. Das, S. K., and Fujimura, R. K. (1979) JBC 254, 1227.
8. McClure, W. R., and Chow, Y. (1980) Meth. Enz. 64, 277; Fay, P. J., Johanson, K. O., McHenry, C. S., and Bambara, R. A. (1981) JBC 256, 976; (1982) JBC 257, 5692.
9. Fay, P. J., Johanson, K. O., McHenry, C. S., and Bambara, R. A. (1982) JBC 257, 5692; Huang, C.-C., Hearst, J. E., and Alberts, B. M. (1981) JBC 256, 4087.
10. Matson, S. W., and Bambara, R. A. (1981) J. Bact, 146, 275.

TABLE S4-1
Processivity of various DNA polymerases[a]

| Source | Polymerase | Template | Processivity value[b] |
|--------|-----------|----------|-----------------------|
| E. coli | pol I | nicked ColE1 | 15–20 |
| | | gapped ColE1 | 45–55 |
| | | oligo dT · poly dA | 11–13 |
| | | poly d(AT) | 170–200 |
| | polA5 | nicked ColE1 | 3–5 |
| | pol III core | primed SS circle[c] | 10–15 |
| | pol III′ | primed SS circle | 40 |
| | pol III* | primed SS circle | 190 |
| | pol III holoenzyme | primed SS circle | >5000 |
| Phage T4 | T4 polymerase[d] | oligo dT · poly dA | 11–13 |
| Phage T5 | T5 polymerase | oligo dT · poly dA | 155–170 |
| Calf thymus | polymerase $\alpha$ | oligo dT · poly dA | 7–9 |
| | polymerase $\beta$ | oligo dT · poly dA | 9–11 |
| KB cell | polymerase $\alpha$ | nicked-gapped calf thymus | 6–16 |
| | polymerase $\beta$ | nicked-gapped calf thymus | 1–2 |
| AMV (virus) | viral polymerase | nicked ColE1 | 22–30 |

[a]Adapted from data supplied by Professor R. A. Bambara.
[b]Nucleotide residues; measured at 37°C, ionic strength of 0.1 $\mu$
[c]Primed phage single-stranded circle
[d]With accessory proteins on a primed SS circle, the processivity value is >20,000 (Section S5-5).

## S4-9 Pyrophosphorolysis and Pyrophosphate Exchange

The more rapid rate of pyrophosphate (PP$_i$) exchange compared to pyrophosphorolysis previously ascribed to action at a transition state (p. 126) may be interpreted in an alternative way.[11] After the diester bond is formed at site A on the template, the polymerization step is completed by movement of the enzyme to the next site (site B) for binding the dNTP for the next addition. Assuming site A of diester bond formation to be the locus of PP$_i$ reactions and site B to be unreactive with PP$_i$, the slower rate of pyrophosphorolysis might be attributed to a rate-limiting step in moving the enzyme back from site B to site A.

Consistent with this interpretation are studies[12] of the PP$_i$ exchange reaction using dATP$\alpha$S as a stereochemical probe. Only the

11. Burgers, P. M. J., personal communication; Krakow, J. S., and Fronk, E. (1969) *JBC 244*, 5988.
12. Marlier, J., Henrie, R., and Benkovic, S. J., personal communication.

diastereomer with the *Sp* configuration was active and after exchange with $^{32}$P-PP$_i$, the radioactively labeled dATPαS retained the *Sp* configuration. These results confirm that PP$_i$ exchange occurs at the stage of diester bond formation and not with a free nucleoside triphosphate.

## S4-10   The 3′→5′ Exonuclease: Proofreading

## S4-11   Polymerase as Mutator or Antimutator

The importance of the 3′→5′ exonuclease activity for fidelity of replication has been demonstrated in additional ways.

(i)   Lack of fidelity (mutagenicity) increases at a given site when the concentration of the next nucleotide to be incorporated is greatly elevated,[13] thereby favoring the step that would prevent removal of a mismatched nucleotide inserted at the given site. Mutagenicity is also favored by high concentrations of dNMPs, which compete with 3′→5′ exonuclease activity. These effects, observed with DNA polymerase I and T4 DNA polymerase but not with a eukaryotic viral DNA polymerase (AMV) that lacks the 3′→5′ exonuclease activity, are interpreted as interfering with proofreading.

(ii)   A deoxynucleoside triphosphate containing a sulfur atom in place of oxygen on the α-phosphate is incorporated into DNA by DNA polymerase I at a rate similar to that of the natural substrate, but DNA terminated with such an α-phosphorothioate is insensitive to the 3′→5′ exonuclease activity of the polymerase. Infidelity in replication by DNA polymerase I or T4 DNA polymerase, measured by insertion of C opposite A, was increased by at least 20-fold with an α-phosphorothioate compared to a normal substrate.[14] With eukaryotic polymerases that lack the 3′→5′ exonuclease activity (e.g., DNA polymerase β, avian myeloblastosis virus DNA polymerase), the use of the α-phosphorothioate analog had no effect on fidelity.

(iii)   Net incorporation of 2-aminopurine (AP) deoxynucleoside triphosphate in competition with dATP by various polymerases is inversely related to the level of 3′→5′ exonuclease activity. In the absence of proofreading, as observed with eukaryotic polymerases and more recently in HeLa cell nuclei,[15] the sevenfold advantage in the incorporation of A over AP can be attributed entirely to the

13. Kunkel, T. A., Schaaper, R. M., Beckman, R. A., and      b, L. A. (1981) *JBC* 256, 9883.
14. Kunkel, T. A., Eckstein, F., Mildvan, A. S., Koplitz, R. M., and Loeb, L. A. (1981) *PNAS* 78, 6734.
15. Wang, M.-L. J., Stellwagen, R. H., and Goodman, M. F. (1981) *JBC* 256, 7097.

1.1-kcal/mol difference in the free energy between an A · T and an AP · T base pair. Nearest-neighbor interactions influence AP incorporation[16] and may explain the "hot spots" observed in early mutation studies.

The role of the 3'→5' exonuclease in repair of DNA lesions is supported by its effectiveness in removing 3'-terminal deoxyribose at incised apurinic-apyrimidinic sites (Section S10-8).

## S4-15  De Novo Synthesis of Repetitive DNA

The apparently unprimed and template-independent synthesis of repetitive homopolymers and copolymers by DNA polymerase I has been attributed to oligonucleotide impurities in the enzyme (p. 146). The claim[17] that such oligonucleotides are made by a contaminating transferase that can polymerize dNDPs without primer or template direction needs to be supported and confirmed.

## S4-18  Physiological Role: Replication and Repair

The initial report that DNA polymerase I (pol I) is absent from viable polA mutants and the numerous demonstrations of its prominent role in DNA repair have obscured in most minds the essential functions of this polymerase in chromosome replication (p. 166). As illustrated by the behavior of various mutants, pol I is needed to remove the 5'-RNA that primes the nascent fragments and to fill the gaps between fragments in the discontinuous phase of replication. These excision and gap-filling capacities of the enzyme have been shown to be essential not only in repair and replication of the host chromosome but also in replication of certain plasmids (Section S14-9). These properties of pol I and its abundance in the cell suggest that it may function as well in a variety of recombinational events, such as transposition of DNA segments.

Transposable elements (e.g., Tn3, Tn5) move to new locations on a genome by a mechanism (Section S16-4) involving duplication of a few nucleotides of the receptor-insertion site. The contribution of pol I to replicative repair in Tn5 transposition is indicated by the

16. Pless, R. C., Levitt, L. M., and Bessman, M. J. (1981) B. 20, 6235.
17. Nazarenko, I. A., Potapov, V. A., Romashchenko, A. G., and Salganik, R. I. (1978) FEBS Lett. 86, 201; Nazarenko, I. A., Bobko, L. E., Romashchenko, A. G., Khripin, Y. L., and Salganik, R. I. (1979) N. A. Res. 6, 2545.

lack of transposition in *polA*-deficient cells.[18] Mutants originally identified by an inability to support transposition of Tn3 were mapped in the *polA* gene with deficiencies in both polymerase and $5' \rightarrow 3'$ exonuclease activities,[19] showing the requirement for both functions. However, the severity of the defects in transposition was not strictly correlated with the degree of deficiency in polymerase and $5' \rightarrow 3'$ exonuclease activities.

An altered form of pol I from cells induced for *rec A-lex A* functions (SOS repair; Section S16-4) shows low fidelity replication in vitro, suggestive of a role in mutagenic repair.[20]

18. Sasakawa, C., Uno, Y., and Yoshikawa, M. (1981) *MGG 182*, 19.
19. Syvanen, M., personal communication.
20. Lackey, D., Krauss, S. W., and Linn, S. (1982) *PNAS 79*, 330.

# S5

# Other Prokaryotic DNA Polymerases

## Abstract

A large polypeptide is the polymerizing subunit in a variety of prokaryotic polymerases (and in the eukaryotic enzymes, Section S6-1). Augmenting the polymerase core with subunits endows it with salt insensitivity, template specificity, high processivity, and catalytic efficiency approaching in vivo values. Discovery of the genetic loci for four of the eight or more subunits of DNA polymerase III holoenzyme of E. coli makes it easier to examine their structures and functions and to label and prepare them in large quantities for such studies. Stoichiometries of the subunits suggest a novel supraholoenzyme at the replication fork (Section S11-14). The T4-phage encoded polymerase with its three auxiliary proteins resembles the host holoenzyme in form and performance. To achieve a more modest replicative mission, the smaller and simpler T7-phage polymerase borrows only host thioredoxin, for some still obscure reason.

## S5-3 DNA Polymerase III Holoenzyme of *E. coli*[1]

### Subunit Structure

Pol III, the holoenzyme core, purified nearly 30,000-fold to virtual homogeneity[2] contains $\alpha$, $\epsilon$, and $\theta$ subunits of 140, 25, and 10 kdal, respectively. The $\alpha$ subunit is the *polC* (*dnaE*) gene product.[3] Resolution of the subunits by gel electrophoresis has not been achieved under native conditions, but polymerase activity assays of proteins resolved under denaturing conditions nonetheless establish the $\alpha$ subunit as the catalytic component[4] (Section S6-1).

Holoenzyme preparations judged to be 50 percent pure or better contain as many as thirteen discrete polypeptides.[5] In addition to those of the core they include $\beta$, $\gamma$, $\delta$, and $\tau$ (Table S5-1). The $\beta$ subunit, isolated as a dimer of 37-kdal polypeptides,[6] is the product of the *dnaN* gene and has a cellular abundance ten times that of the holoenzyme.[7] Transcription of the *dnaN* gene proceeds from the promoter of the adjacent *dnaA* gene whose protein product is essential for initiation of replication of the *E. coli* chromosome. In this way elongation of replication may be regulated by the initiation system.

The $\gamma$ subunit (52 kdal) is the product of the *dnaZ* gene;[8] the $\delta$ subunit (32 kdal) is encoded by *dnaX*.[9] Proximity of the *dnaZ* and *dnaX* genes makes it possible that both are present in a plasmid that leads to dnaZ protein ($\gamma$ subunit) overproduction[8]. Were dnaX protein ($\delta$ subunit) also overproduced in cells with this plasmid, then overproduction of a $\gamma \cdot \delta$ complex in this strain (p. 177) could be explained.

1. McHenry, C., and Kornberg, A. (1981) *Enzymes 14*, 39.
2. McHenry, C., and Crow, W. (1979) *JBC 254*, 1748.
3. McHenry, C., and Welch, M., personal communication.
4. Spanos, A., Sedgwick, S. G., Yarronton, G. T., Hübscher, U., and Banks, G. R. (1981) *N. A. Res. 9*, 1825.
5. McHenry, C. (1980) *ICN–UCLA Symposia 19*, 569; (1982) *JBC 257*, 2657.
6. Johanson, K. O., and McHenry, C. S. (1980) *JBC 255*, 10984; McHenry, C. S., personal communication.
7. Burgers, P. M. J., Kornberg, A., and Sakakibara, Y. (1980) *PNAS 78*, 5391.
8. Hübscher, U., and Kornberg, A. (1980) *JBC 255*, 11698.
9. Hübscher, U., and Kornberg, A. (1979) *PNAS 76*, 6284.

TABLE S5-1
Components of DNA polymerase III holoenzyme

| Subunit | Mass, kdal | Alternative designations |
|---------|------------|--------------------------|
| $\alpha$ | 140 | dnaE protein, polC protein |
| $\epsilon$ | 25 | — |
| $\theta$ | 10 | — |
| $\tau$ | 83 | — |
| $\gamma$ | 52 | dnaZ protein |
| $\delta$ | 32 | Factor III, dnaX protein |
| $\beta$ | 37 | Factor I, copol III*, dnaN protein |

(Bracketing: $\alpha$, $\epsilon$, $\theta$ = pol III (core); core + $\tau$ = pol III′; through $\delta$ = pol III*)

## Pol III′

Pol III′, a new form of DNA polymerase III, has been purified 15,000-fold to 90 percent homogeneity.[10] It is a subassembly of four subunits of the holoenzyme: $\tau$ (83 kdal) and the three core subunits, $\alpha$, $\epsilon$, and $\theta$. Molecular weight estimates suggest an entity with two units of core and two of $\tau$. The functional and physical properties of pol III′ are intermediate between pol III and the holoenzyme. Pol III′ is clearly distinguished from pol III by its capacity to utilize a long single-stranded template in the presence of spermidine, but unlike the holoenzyme it is virtually inert on such a template coated with single-strand binding protein.

An 83-kdal polypeptide only slightly larger than $\tau$ with potent DNA-dependent ATPase activity is generally found in holoenzyme preparations,[11] but can be removed apparently without affecting their function.

## Mechanism

In vitro, activation of a polymerase to form an initiation complex with a primer-template is more than a thousandfold slower than the polymerization step. For this reason, replication of extensive lengths of template is aided by high processivity (p. 121). Pol III, with a processivity of 10 to 15 residues, is effective only in filling short gaps. Pol III holoenzyme by contrast has apparently unlimited processivity (>5000 residues). Pol III′ and pol III* are intermediate with respective processivity values of 40 and 190 residues (Table S4-1).[12]

10. McHenry, C. S. (1982) *JBC 257*, 2657.
11. Meyer, R. R., Brown, C. L., and Rein, D. C., personal communication.
12. Fay, P. J., Johanson, K. O., McHenry, C. S., and Bambara, R. A. (1981) *JBC 256*, 976; (1982) *JBC 257*, 5692.

Pol III holoenzyme action is distinguished from the behavior of incomplete subunit assemblies such as pol III, pol III', and pol III* (Table S5-1) by its absolute dependence on ATP (or dATP) activation and relative insensitivity to the reduction in processivity caused by high ionic strength.[13] All other nucleoside triphosphates, including nonhydrolyzable ATP analogs, failed to substitute for ATP. ATP binds tightly to holoenzyme ($K_D = 0.8$ $\mu$M), but only feebly ($K_D \geqq 5$ $\mu$M) to the holoenzyme lacking the subunit (pol III*). Binding of ATP to the free $\beta$ subunit was not detectable, but reconstitution of the holoenzyme from pol III* and $\beta$ subunit restored tight binding ($K_D = 0.6$ $\mu$M). Approximately two or three ATPs were bound per replicative unit of holoenzyme without apparent cooperativity. The ATP-holoenzyme complex is required in the formation of an initiation complex with a primed template but not for subsequent DNA synthesis. The nucleotide analog, dAMPPNP (2'-deoxadenyl-5'-imidodiphosphate) can substitute for dATP in chain elongation but not in holoenzyme activation.

The ATPase activity of the holoenzyme is specifically elicited by template-primer DNA. Reaction of a preformed ATP · holoenzyme complex with a primed template results in hydrolysis of the ATP bound to the holoenzyme, release of ADP and $P_i$, and formation of an initiation complex between holoenzyme and the primed template. Approximately two ATPs are hydrolyzed for each initiation complex formed, a value in keeping with the number of ATPs bound in the ATP · holoenzyme complex. The possibility that the latter and the initiation complex contain two holoenzyme molecules is supported by the presence of two molecules of the $\beta$ subunit in the initiation complex. Holoenzyme actions in the absence of ATP resemble those of pol III or pol III* with or without ATP in sensitivity to salt and in processivity of elongation. The initiation complex formed by ATP-activated holoenzyme resists a level of KCl (150 mM) that completely inhibits unactivated holoenzyme and the incomplete forms of the holoenzyme, and displays a processivity at least 20 times greater. Upon completing replication of available template, holoenzyme can dissociate and form an initiation complex with another primed template, provided ATP is available to reactivate the holoenzyme; no essential subunits are lost in the cycle of initiation, elongation, and dissociation.[13] Failure of an *anti-β* subunit antibody to inhibit elongation by the initiation complex is not due to loss of the $\beta$ subunit from the complex but rather to the inaccessibility of this subunit in the complex.[14]

---

13. Burgers, P. M. J., and Kornberg, A., unpublished results.
14. Johanson, K. O., and McHenry, C. S., personal communication.

The rate of DNA synthesis by pol III holoenzyme on single-stranded phage DNA coated with single-strand binding protein is 700 nucleotides/sec at 37°C, a value approaching the in vivo rate of fork movement of 1000 nucleotides/sec. Thus the holoenzyme may operate in vitro nearly as well as it does in the cell. Stoichiometries of several of the holoenzyme subunits and ATP suggest the existence of a twin holoenzyme assembly and with it a mechanism for carrying out simultaneous or concurrent elongation of both leading and lagging strands at the replication fork (Section S11-14).

## S5-5 Phage T4-Encoded DNA Polymerase[15]

Accessory phage-encoded proteins are akin to the pol III holoenzyme subunits in facilitating replication by the polymerase subunit. They enable the gene 43 polymerase to replicate a single-stranded template at an increased rate and permit the efficient utilization of an otherwise inaccessible nicked duplex DNA. The functions of these proteins have been clarified by two model systems.[16]

In the first system, with a primed, single-stranded, circular DNA,[17] the accessory proteins (encoded by genes 44, 62, and 45; see Table 14-12, p. 534) increase the rate fourfold; gene 32 protein (p. 282) increases the rate sixfold; with the combined action of these proteins in a 5-component system (43, 44, 62, 45, and 32), the rate is increased more than fortyfold. The accessory proteins seem to clamp the polymerase to the template, thereby enabling the enzyme to traverse and replicate the strongest hairpin helices in a highly processive manner (>20,000 nucleotides). While ATP hydrolysis is needed to form the holoenzyme-like complex of polymerase and accessory proteins, the nonhydrolyzable analog, ATPγS, is thereafter able to sustain elongation.

In the second model system,[18] the polymerase starts replication at a nick in an otherwise intact duplex DNA. The rate of the 5-protein system is only 20 percent of the estimated rate of fork movement in vivo. Increasing the gene 32 protein concentration to a great excess over that needed to bind the displaced single strand increases fork movement by what may be a helicase action. A threefold increase in rate of fork movement results from the overt helicase

15. Lehman, I. R. (1981) *Enzymes 14*, 51.
16. Alberts, B. M., Barry, J., Bedinger, P., Burke, R. L., Hibner, U., Liu, C.-C., and Sheridan, R. (1980) *ICN–UCLA Symposia 19*, 449.
17. Huang, C.-C., Hearst, J. E., and Alberts, B. M. (1981) *JBC 256*, 4087; Roth, A. C., Nossal, N. G., and Englund, P. T. (1982) *JBC 257*, 1267.
18. Alberts, B. M., Barry, J., Bedinger, P., Burke, R. L., Hibner, U., Liu, C.-C., and Sheridan, R. (1980) *ICN–UCLA Symposia 19*, 449.

activity of gene 41 protein (Section S9-9); rates are not increased by the presence or priming action of gene 61 protein (Section 11-10).

The high degree of base-pairing fidelity of the T4 multienzyme complex is described in Section S11-12.

## S5-6 Phage T5-, T7-, and SPO1-Encoded DNA Polymerases[19]

### Phage T5

A transcription-replication complex has been described[20] with *E. coli* RNA polymerase, T5 DNA polymerase, DNA-binding protein, transcriptional factors, and the gpD15 nuclease among its components.

A kinetic study of T5 DNA polymerase[21] demonstrated high processivity both for polymerization and for $3' \to 5'$ exonuclease action; elongation and turnover at a 3'-OH primer terminus can be catalyzed by the same enzyme molecule.

### Phage T7

Gene 5 of phage T7 encodes an 84-kdal protein, which together with host-cell thioredoxin (12 kdal) constitutes the DNA polymerase. The function of thioredoxin is still unknown. The gene 5 protein alone is a $3' \to 5'$ single-stranded DNA exonuclease with activity on both single- and double-stranded DNA.[22] T7 DNA polymerase can exist in two forms.[23] When purified in the presence of EDTA, the polymerase has elevated levels of exonuclease activities on both single- and double-stranded DNA and cannot initiate synthesis at nicks either in the presence or absence of T7 gene 4 protein. When purified in the absence of EDTA, the polymerase catalyzes limited synthesis at nicks in duplex DNA and is markedly stimulated by T7 gene 4 protein. The ability to be stimulated by gene 4 protein most likely reflects the displacement of a single strand to which the gene 4 protein can bind and synthesize a transcript for priming (Section S11-10). No physical difference between the two forms of polymerase has yet been shown.

19. Lehman, I. R. (1981) *Enzymes 14*, 51.
20. Ficht, T. A., and Moyer, R. W. (1980) *JBC 255*, 7040.
21. Das, S. K., and Fujimura, R. K. (1980) *JBC 255*, 7149.
22. Hori, K., Mark, D. F., and Richardson, C. C. (1979) *JBC 254*, 11591; 11598; Adler, S., and Modrich, P. (1979) *JBC 254*, 11605.
23. Fischer, H., and Hinkle, D. C. (1980) *JBC 255*, 7956; Engler, M. J., Lechner, R. L., and Richardson, C. C., personal communication.

# S6

# Eukaryotic
# DNA Polymerases

## Abstract

The controversies that swirled over the cellular location, size, and subunit organization of DNA polymerase $\alpha$ have now been resolved. This principal enzyme of chromosome replication is found localized to the nucleus, where it belongs, when detected with a proper immunocytologic technique. In all the replicative eukaryotic DNA polymerases the polymerizing activity resides in a large polypeptide (in the 125-kdal range). A variety of smaller, active polypeptides, some only 35 kdal in size, are the result of proteolysis during the course of isolation. The large polypeptide is the core of a holoenzyme form containing several more subunits, which together confer processivity and thus enable the enzyme to replicate long stretches of single-stranded template efficiently.

Detailed kinetic analysis of the polymerase $\alpha$ mechanism may have wide significance for polymerase actions. The enzyme reacts first with the template, then with the primer, and finally, conformationally altered by the template, binds a specific deoxyribonucleoside triphosphate. The latter is chelated by a divalent metal cation (e.g., $Mg^{2+}$) that may also serve in the catalytic and translocation stages of each replicative event.

More has also been learned about the physical and catalytic features of DNA polymerase $\beta$. The enzyme structure is highly conserved from birds to humans, but little is yet known of its true functions. In addition to polymerases $\alpha$, $\beta$, and $\gamma$, a fourth mammalian DNA polymerase, named $\delta$, has been discovered, which is unique in its intrinsic $3' \rightarrow 5'$ exonuclease with "editing" properties like those of prokaryotic polymerases.

Terminal deoxynucleotidyl transferase, whose contribution to lymphocyte maturation is still enigmatic, is now established as a 60-kdal polypeptide, widely shared by animals, rather than the proteolytic fragments that heretofore characterized it.

## S6-1 Animal DNA Polymerases[1]

Controversies concerning the size and cellular location of polymerase $\alpha$ have been resolved. Catalytic activity resides in a large subunit of 120 to 200 kdal (Section S6-2). The nuclear location of DNA polymerase $\alpha$ has been demonstrated in a more definitive way with murine monoclonal antibodies against the enzyme from a human malignant KB cell line (Section S6-2). In KB cells fixed with paraformaldehyde and monitored with a peroxidase-coupled, anti-mouse IgG detection system the $\alpha$ polymerase was clearly and exclusively in the nucleus.[2] A contrary claim[3] was based on a procedure with a methanol fixative that likely permitted extraction and redistribution of soluble cellular components.

### DNA Polymerase $\delta$

A new mammalian DNA polymerase, named $\delta$, has been obtained from calf thymus and from erythroid hyperplastic rabbit bone marrow.[4] Delta is unique among DNA polymerases of higher eukaryotes in having an intrinsic $3' \rightarrow 5'$ exonuclease activity with "editing" properties similar to those of *E. coli* DNA polymerase I (p. 127). The $\delta$ polymerase from calf thymus has a molecular weight of 140 to 160 kdal and is composed of two dissimilar subunits. Like DNA polymer-

1. Scovassi, A. I., Plevani, P., and Bertazzoni, U. (1980) *TIBS 5*, 335; De Pamphilis, M. L., and Wassarman, P. M. (1980) *ARB 49*, 627; Weissbach, A. (1981) *Enzymes 14*, 67.
2. Bensch, K., Tanaka, S., Hu, S., Wang, T. S-F., and Korn, D., personal communication.
3. Brown, M., Bollum, F. J., and Chang, L. M. S. (1981) *PNAS 78*, 3049.
4. Tsang Lee, M. Y. W., Tan, C.-K., So, A. G., and Downey, K. M. (1980) *B. 19*, 2096; Goscin, L. P., and Byrnes, J. J. (1982) *B.*, in press.

ase α, the calf thymus δ polymerase is inhibited by N-ethylmaleimide, aphidicolin, and araATP, and is active in the presence of dideoxy-TTP.[5] Its role in DNA replication in vivo is not known. A similar enzyme, termed DNA polymerase $\alpha_1$, has been isolated from mouse myeloma cells.[6]

## Inhibitors

For lack of polymerase mutants, great reliance is placed on inhibitors to characterize animal DNA polymerases in vitro and in vivo. Conclusions are strengthened when concordant results are obtained with several inhibitors (Table S6-1).

TABLE S6-1
DNA polymerase inhibitors[a]

| Inhibitor | Inhibition, percent | | |
|---|---|---|---|
| | α | β | γ |
| NEM (N-ethylmaleimide) (1 mM) | 100 | 0 | 85 |
| ddTTP (ddTTP/dTTP = 10) | 10 | 90 | 90 |
| Aphidicolin (2 μg/ml) | 90 | 0 | 0 |
| araCTP (125 μM) | 85 | 15 | |
| Butylanilinouracil (100 mM) | 60 | 0 | 0 |
| β-Lapachone (83 mM) | 60 | 10 | |
| Pyridoxal-5-phosphate (0.4 mM) | 96 | 55 | 92 |
| Heat inactivation (10 min, 45°C) | 0 | 90 | |
| Heparin (0.2 μg/ml) | 70 | 10 | 10 |
| Ethidium bromide (20 μM) | 10 | 10 | 60 |

[a]From Scovassi, A. I., Plevani, P., and Bertazzoni, U. (1980) *TIBS* 5, 335.

# S6-2   DNA Polymerase α

The role of DNA polymerase α as the principal replicating enzyme is further established by the specific inhibitory action of aphidicolin, as useful a guide for eukaryotic replication as rifampicin for prokaryotic transcription. Isolation of an aphidicolin-resistant DNA

5. Tsang Lee, M. Y. W., Tan, C.-K., So, A. G. and Downey, K. M. (1980) *B.* 19, 2096.
6. Chen, Y.-C., Bohn, E. W., Planck, S. R., and Wilson, S. H. (1979) *JBC* 254, 11678.

polymerase α mutant of *Drosophila,* nicely correlated with replication in vivo, is an interesting example.[7] An additional function for the animal cell polymerase in repair of DNA lesions seems likely.[8]

An α-like DNA polymerase with characteristic aphidicolin sensitivity[9,10] has been identified in spinach leaves and cell suspension cultures of rice. The enzyme is clearly distinguished from a γ-like DNA polymerase purified from chloroplasts (Section S6-4).

Size and Subunit Organization

The nearly homogeneous DNA polymerase α from *Drosophila melanogaster* embryos has a sedimentation coefficient of 9S and a molecular weight of 280,000. It is composed of four subunits:[11] α, 148 kdal; β, 58 kdal; γ, 46 kdal; and δ, 43 kdal. In this respect it resembles the comparable enzyme from regenerating rat liver, which contains five subunits[12] with sizes of 156, 64, 61, 58, and 54 kdal. In both enzymes, DNA polymerase activity, measured with activated DNA as template-primer, is associated with the large (148 or 156 kdal) subunit. Tryptic digests of the catalytic subunits of the *Drosophila* and rat liver enzymes have eight to ten peptides in common, suggesting a significant homology,[13] although comparable immunological evidence is lacking.[14] Calf thymus polymerase α, previously reported as lacking a large subunit, has now been purified to near homogeneity as a 9S holoenzyme (mol. wt. 500,000) composed of subunits[15] of 148, 59, 55, and 48 kdal.

DNA polymerase α activity in extracts of a wide variety of eukaryotic cells has been identified in situ in a polyacrylamide gel after electrophoresis under denaturing conditions.[16] In every instance, the catalytic activity was observed in a polypeptide of 120 to 200 kdal. After a brief incubation of the crude enzyme fractions, the large polypeptide was reduced to smaller forms that were still catalytically active.[17]

In view of all the evidence, it seems likely that the polypeptides

7. Sugino, A., and Nakayama, K. (1980) *PNAS 77,* 7049.
8. Miller, M. R., and Chinault, D. N. (1982) *JBC 257,* 46.
9. Amileni, A., Sala, F., Cella, R., and Spadari, S. (1979) *Planta 146,* 521; Sala, F., Parisi, B., Burroni, D., Amileni, A., Pedrali-Noy, G., and Spadari, S. (1980) *FEBS Lett. 117,* 93.
10. Sala, F., Amileni, A., Parisi, B., and Spadari, S. (1980) *EJB 112,* 211.
11. Banks, G. R., Boezi, J. A., and Lehman, I. R. (1979) *JBC 254,* 9886; Villani, G., Sauer, B., and Lehman, I. R. (1980) *JBC 255,* 9479.
12. Méchali, M., Abadiedebat, J., and de Recondo, A.-M. (1980) *JBC 255,* 2114.
13. Méchali, M., and de Recondo, A.-M., personal communication.
14. Chang, L. M. S., and Bollum, F. J. (1981) *JBC 256,* 494.
15. Grosse, F., and Krauss, G. (1981) *B. 20,* 5470.
16. Spanos, A., Sedgwick, S. G., Yarronton, G. T., Hübscher, U., and Banks, G. R. (1981) *N. A. Res. 9,* 1825; Albert, W., Grummt, F. Hübscher, U., and Wilson, S. H. (1982) *N. A. Res. 10,* 935.
17. Hübscher, U., Spanos, A., Albert, W., Grummt, F., and Banks, G. R. (1981) *PNAS 78,* 6771.

of highest molecular weight represent the valid, highly conserved catalytic subunits of the $\alpha$ DNA polymerases. Moreover, when an active low-molecular-weight polypeptide bears the polymerase $\alpha$ activity, as reported for enzymes purified from KB cells[18] and mouse myeloma,[19] it seems likely that the polypeptide is the product of proteolysis of a larger form due to an abundance of an endogenous protease or to a special sensitivity of the polymerase to proteolysis. Similar concerns about proteolysis extend to the accessory subunits of the holoenzyme form of the polymerase.

Accessory Subunits

Although the large $\alpha$ subunit of the *Drosophila* $\alpha$ polymerase is able to replicate duplex DNA with short single-stranded gaps, the subunit is incapable of replicating RNA-primed, single-stranded phage M13 DNA, a reaction that the intact "holoenzyme" can catalyze. Accessory proteins for replication of long lengths of single-stranded template are also implicated in the actions of HeLa cell polymerase $\alpha$ (140 kdal), from which two distinctive native proteins of 96 and 51 kdal were resolved; the 96-kdal protein appeared to be made up of still smaller subunits.[20] The accessory proteins do not enhance the activity of the polymerase on the standard "activated" template (with nicks and small gaps) but are required for full activity on a single-stranded template, on which the polymerase alone is inert. For extensive replication of an RNA-primed, single-stranded phage $\phi$X174 DNA, DNA polymerase $\beta$ was able to complement the action of polymerase $\alpha$.[21]

Another interesting accessory protein is an $A_{P_4}A$ (ApppppA) binding protein tightly associated with a 640-kdal form of HeLa cell DNA polymerase $\alpha$.[22] $A_{P_4}A$, proposed as a positive growth signal (p. 468), can prime synthesis of poly dA on a poly dT template with this polymerase complex.

In its relationship to accessory proteins, DNA polymerase $\alpha$, the enzyme largely responsible for chromosome replication, resembles the core and holoenzyme forms of *E. coli* DNA polymerase III (Section S5-3). Greater processivity and catalytic efficiency for long stretches of single-stranded template depend on the presence of multiple subunits in a truly native DNA polymerase $\alpha$ holoenzyme.[23]

18. Fisher, P. A., and Korn, D. (1977) *JBC 252*, 6528; Filpula, D., Fisher, P. A., and Korn, D. (1982) *JBC 257*, 2029.
19. Chen, Y.-C., Bohn, E. W., Planck, S. R., and Wilson, S. H. (1979) *JBC 254*, 11678.
20. Lamothe, P., Baril, B., Chi, A., Lee, L., and Baril, E. (1981) *PNAS 78*, 4723.
21. Ikeda, J.-E., Longiaru, M., Horwitz, M. S., and Hurwitz, J. (1980) *PNAS 77*, 5827.
22. Rapaport, E., Zamecnik, P. C., and Baril, E. F. (1981) *JBC 256*, 12148.
23. Villani, G., Fay, P. J., Bambara, R. A., and Lehman, I. R. (1981) *JBC 256*, 8202.

A novel and clarifying finding[24] is the presence of a potent decanucleotide primase activity (Section S11-10) in the purified *Drosophila* DNA polymerase α preparation. This is the first example of a eukaryotic primase and appears to be an intrinsic component of the holoenzyme. Primase activity has also been found in the purified DNA polymerase α of rat liver.[25] These complexes of DNA polymerase α and primase may prove to be a major step toward the recognition and capture of the elusive eukaryotic replisome.

## Mechanism

Insights of general significance regarding the catalytic mechanism and conformational states of a DNA polymerase come from kinetic and direct-binding assays of near-homogeneous human DNA polymerase α.[26] The enzyme reacts with its substrates in an ordered sequential mechanism, with template (single-stranded DNA) as the first substrate, followed by primer, and then by dNTP. Which dNTP adds to the enzyme is influenced by template sequence, presumably the result of conformational alteration of the enzyme by the binding of template. Each catalytically active polymerase α molecule appears to possess at least two strongly interactive template sites, suggestive of two active centers. Primer binding requires a primer of at least eight nucleotides with the terminal three to five nucleotides basepaired. Mispairing of the primer terminus prevents correct primer binding as well as dNTP addition to the polymerase.

$Mg^{2+}$ is a potent effector of the primer-binding reaction. The interaction involves a $Mg^{2+}$-primer chelate and the coordinated participation of four $Mg^{2+}$-primer-binding subsites. In a proposed model, $Mg^{2+}$ is brought to the polymerase-template-primer complex by dNTPs and is then passed vectorially and processively to each of the four postulated $Mg^{2+}$-primer-binding subsites, thereby facilitating both phosphodiester bond formation and polymerase translocation. These reactions could explain how $Mn^{2+}$ and other divalent metals reduce the fidelity of DNA replication (p. 406). Kinetic data[27] on the action of a mouse myeloma polymerase α are also in keeping with an ordered sequential mechanism and two distinct template-primer binding sites (effector and catalytic) on each enzyme molecule.

24. Conaway, R., and Lehman, I. R. (1982) *PNAS 79*, 2523.
25. de Recondo, A.-M., personal communication.
26. Fisher, P. A., and Korn, D. (1977) *JBC 252*, 6528; (1979) *JBC 254*, 11033, 11040; (1981) *B. 20*, 4560, 4570; Fisher, P. A., Chen, J. T., and Korn, D. (1981) *JBC 256*, 133.
27. Detera, S. D., Becerra, S. P., Swack, J. A., and Wilson, S. H. (1981) *JBC 256*, 6933.

DNA polymerase $\beta$, purified from chick embryos to apparent homogeneity,[28] is a single polypeptide of 40 kdal; similar results were obtained with mouse myeloma, rat liver and hepatoma, rabbit and pig liver, and calf thymus as sources.[29] Tryptic peptide maps of radioiodinated proteins showed extensive homology (>80 percent) among the polypeptides obtained from all of these mammalian species, as well as between them and the chick embryo enzyme. The biochemical data agree with immunological results[30] in which a polyvalent antiserum against calf thymus polymerase $\beta$ cross-reacted with polymerase $\beta$ fractions from mammals and chick embryos but only feebly with a toadfish enzyme, and not at all with a protozoan polymerase $\beta$, nor with other eukaryotic polymerases from any source. Clearly, DNA polymerase $\beta$ is highly conserved, with some indications for a role in DNA repair shared with DNA polymerase $\alpha$.[31]

### Mechanism

Kinetic studies of DNA polymerase $\beta$ purified from myeloma[32] and human liver or KB cells[33] confirm that polymerization is almost exclusively distributive. The reaction mechanism is ordered and sequential,[34] with DNA binding first, followed by a dNTP. There is limited, strand-displacement synthesis on nicked, duplex DNA substrates. Reaction velocities are maximal on DNA primer-templates containing short gaps of ten to twenty nucleotides.[35] The affinity of the enzyme for nicked duplex DNA molecules is extremely high, but feeble or absent for single-stranded or intact duplex DNA. Thus, a base-paired primer adjacent to a very short single-stranded region (actual or potential) is the preferred template-primer.

With $Mg^{2+}$, the minimum template length is five nucleotides, whereas a template of a single nucleotide suffices with $Mn^{2+}$. The divalent cation, previously shown[36] to affect the substrate $K_m$, $V_{max}$,

28. Yamaguchi, M., Tanabe, K., Taguchi, Y. N., Nishizawa, M., Takahashi, T., and Matsukage, A. (1980) *JBC 255,* 9942.
29. Tanabe, K., Yamaguchi, M., Matsukage, A., and Takahashi, T. (1981) *JBC 256,* 3098.
30. Chang, L. M. S., and Bollum, F. J. (1981) *JBC 256,* 494.
31. Miller, M. R., and Chinault, D. N. (1982) *JBC 257,* 46.
32. Tanabe, K., Bohn, E. W., and Wilson, S. H. (1979) *B. 18,* 3401.
33. Wang, T. S.-F., and Korn, D. (1980) *B. 19,* 1782; (1982) *B. 21,* 1597.
34. Wang, T. S.-F., Eichler, D. C., and Korn, D. (1977) *B. 16,* 4927; (1982) *B. 21,* 1597.
35. Wang, T. S.-F., and Korn, D. (1980) *B. 19,* 1782.
36. Wang, T. S.-F., and Korn, D. (1980) *B. 19,* 1782; Wang, T. S.-F., Eichler, D. C., and Korn, D. (1977) *B. 16,* 4927.

and the catalytic efficiency of the human polymerase $\beta$ reaction with DNA, affects the processivity of dNMP incorporation identically on nicked and on gapped DNA primer-templates.[37] The polymerization mechanism with $Mg^{2+}$ is therefore distributive, while with $Mn^{2+}$ it is processive to the extent that four to six nucleotides are inserted in each binding cycle.

## S6-4   DNA Polymerase $\gamma$

### Animals

The mitochondrial enzyme has been purified to near homogeneity from chick embryos,[38] the first time that a polymerase of this class has been purified so extensively from any source. From a native molecular weight near 180,000 and a single major polypeptide of 47 kdal in SDS gel electrophoresis, a tetramer of four identical subunits is suggested. With the primer-template, poly $rA \cdot (dT)_{12-18}$, the enzyme is extremely processive, synthesizing full-length product chains with each binding event.

### Plants

A polymerase purified from spinach chloroplasts[39] shares many properties with the mammalian mitochondrial enzyme: preference for a poly $rA \cdot (dT)_{12-18}$ template, $Mn^{2+}$ optimum of 0.1 to 1 mM, KCl optimum of 100 mM, pH optimum of 8 to 9, large size (105 kdal), resistance to aphidicolin, and sensitivity to N-ethylmaleimide. In view of the resemblance of chloroplast DNA to mitochondrial DNA in its lack of nucleosomal organization and in the presence of displacement loops in its replicative forms, the similarities in physical and functional properties of the plant $\gamma$-like polymerase to the mammalian enzyme are not surprising.

## S6-5   DNA Polymerases of Lower
and Unicellular Eukaryotes[40]

Yeast DNA polymerase I rather than II appears to be the enzyme principally responsible for chromosome replication. Both enzymes are sensitive to aphidicolin, but from a mutant strain selected for

37. Wang, T. S.-F., and Korn, D. (1982) *B. 21*, 1597.
38. Yamaguchi, M., Matsukage, A., and Takahashi, T. (1980) *JBC 255*, 7002; (1980) *Nat. 285*, 45.
39. Sala, F., Amileni, A. R., Parisi, B., and Spadari, S. (1980) *EJB 112*, 211.
40. Scovassi, A. I., Plevani, P., and Bertazzoni, U. (1980) *TIBS 5*, 335.

resistance to aphidicolin, the isolated polymerase I proved to be resistant while polymerase II retained its sensitivity.[41] (Yeast DNA polymerase II may prove to have a role analogous to that of pol I in *E. coli*.) Replication of the 2-$\mu$m yeast plasmid by yeast extracts (Section S15-8) was twenty to forty times more resistant to aphidicolin when extracts were prepared from drug-resistant mutants as compared to wild-type cells. Although this finding implicates polymerase I as the replicase primarily responsible for plasmid replication, the purified enzyme failed to replace any of several enzyme fractions needed to reconstitute the reaction,[42] possibly because a more complex, holoenzyme form is needed.

A phylogenetic survey of the occurrence of the three principal forms of eukaryotic DNA polymerases is presented in Table S6-2. Properties of these polymerases with respect to function, size, associated exonuclease activity, and sensitivity to N-ethylmaleimide are summarized in Table S6-3. The confusion introduced by proteolysis regarding the size and subunit organization of mammalian polymerases (Section S6-3) in all likelihood applies to other eukaryotes as well.

---

41. Sugino, A., Kojo, H., Greenberg, B. D., Brown, P. O., and Kim, K. C. (1981) *ICN–UCLA Symposia* 22, 529.
42. Sugino, A., personal communication.

TABLE S6-2
Phylogeny of eukaryotic DNA polymerases[a]

| Phyla | Classes | $\alpha$-like | $\beta$-like | $\gamma$-like |
|---|---|---|---|---|
| Vertebrates | mammals | + | + | + |
| | birds | + | + | + |
| | reptiles | + | + | + |
| | amphibians | + | + | + |
| | fishes | + | + | + |
| Arthropods | insects | + | +[b] | + |
| Echinoderms | echinoids | + | + | + |
| Mollusks | cephalopods | + | + | |
| Coelenterates | hydrozoans | + | + | |
| Porifera | sponges | + | + | + |
| Protozoa | ciliates | + | − | |
| | flagellates (parasitic) | + | + | |
| Thallophytes | ascomycetes | + | − | + |
| Spermaphytes | monocotyledons | + | − | + |

[a]From Scovassi, A. I., Plevani, P., and Bertazzoni, U. (1980) *TIBS* 5, 335.
[b]Polymerase $\beta$ has been identified in adult insects (*Drosophila*). [Furia, M., Polito, L. C., Locorotoudo, G., and Grippo, P. (1979) *N. A. Res.* 6, 3390; Lehman, I. R., personal communication.]

DNA polymerases from eukaryotes[a]

| Organisms | Type | Function | Mass, kdal | Associated exo | NEM sensitivity |
|---|---|---|---|---|---|
| Animals | α | DNA replication | 130–280 | no | yes |
| | β | DNA repair | 30–50 | no | no |
| | γ | adenovirus DNA replication | 150–300 | no | yes |
| | δ | unknown | 190 | yes | yes |
| | mt-γ | mt-DNA replication | 150–300 | no | yes |
| *Euglena gracilis* | A | unknown | 190 | no | yes |
| | B | unknown | 240 | yes | yes |
| *Tetrahymena pyriformis* | I | DNA replication | 80–130 | yes | yes |
| | II | unknown | 70 | | no |
| *Trypanosoma brucei* | major | unknown | 100 | | yes |
| | minor | unknown | 50 | | no |
| Yeast | I; A | DNA replication | 150 | no | yes |
| | II; B | unknown | 150 | yes | yes |
| *Ustilago maydis* | major | DNA replication | 180 | yes | yes |
| *Neurospora crassa* | A | unknown | 150 | no | |
| | B | unknown | 110 | no | |
| *Dictyostelium discoideum* | single | unknown | 130 | no | yes |
| *Physarum polycephalum* | major | unknown | 120–200 | | yes |

[a]From Scovassi, A. I., Plevani, P., and Bertazzoni, U. (1980) *TIBS 5*, 335.

## S6-7   DNA Virus-Induced DNA Polymerases

Herpes Virus[43]

All members of the herpes virus group (Section S15-5) induce a new polymerase.[44] The level in infected cells may be several times greater than that of all host cell DNA polymerases combined. The purified Herpes Simplex Virus (HSV) polymerase is a polypeptide of about 150 kdal[45] and is strongly inhibited by phosphonacetate, phosphoformate,[46] and acycloguanosine (acyclovir).[47] Aphidicolin, the host DNA polymerase inhibitor, also inhibits HSV polymerase.[48]

43. Knipe, D. M., Ruyechan, W. T., and Roizman, B. (1979) *J. Virol. 29,* 698; Chartrand, P., Crumpacker, C. S., Schaffer, P. A., and Wilkie, N. M. (1980) *Virol. 103,* 311; Crumpacker, C. S., Chartrand, P., Subak-Sharpe, J. H., and Wilkie, N. M. (1980) *Virol. 105,* 171; Purifoy, D. J. M., and Powell, K. L. (1981) *J. Gen. Virol. 54,* 219.
44. Weissbach, A. (1979) *Arch. Biochem. Biophys.* 198, 386.
45. Knopf, K. (1979) *EJB 98,* 231.
46. Eriksson, B., Larsson, A., Helgstrand, E., Johansson, N.-G., and Öberg, B. (1980) *BBA 607,* 53.
47. Furman, P. A., St. Clair, M. H., Fyfe, J. A., Rideout, J. L., Keller, P. M., and Elion, G. B. (1979) *J. Virol. 32,* 72; Miller, W. H., and Miller, R. L. (1980) *JBC 255,* 7204; Darby, G., Field, H. J., and Salisbury, S. A. (1981) *Nat. 289,* 81.
48. Pedrali-Noy, G. and Spadari, S. (1980) *J. Virol. 36,* 457.

Mutations within a 5-kb region of the genome can cause thermo-lability of the enzyme and may also generate resistance to phosphonacetate and acycloguanosine (see below), both in vivo and in vitro. Surprisingly, two distinct complementation groups for the polymerase are found in this small region of the genome, and temperature-sensitive mutations in each yield thermolabile polymerase products. Complexity of polymerase organization and intragenic complementation are among the suggested explanations.

The locus of phosphonacetate inhibition of the herpes polymerase was thought to be the inorganic pyrophosphate binding site of the enzyme (p. 439). However, the concurrence of resistance to phosphonacetate, acycloguanosine, and araA bring this simple mechanism into question. The basis for acycloguanosine action is: (i) its conversion to the monophosphate only by the virus-specific thymidine kinase, and (ii) its inhibition as a nucleoside triphosphate of the viral DNA polymerase which is ten to thirty times more sensitive than the cellular polymerases.[49]

### Vaccinia Virus

The DNA polymerase induced by vaccinia virus, a cytoplasmic pox virus (Section S15-6), has been purified to homogeneity.[50] It is a polypeptide of 110 to 115 kdal with a potent $3' \rightarrow 5'$ exonuclease activity. This polymerase is less able than *E. coli* DNA polymerase I to utilize a primed single-strand template, being stopped readily by secondary structure (e.g., hairpins) in the template.[51] The enzyme, like DNA polymerase $\alpha$ and the herpes-induced polymerase, is inhibited by aphidicolin.[52]

## S6-8  RNA-Directed DNA Polymerases: Viral Reverse Transcriptases[53]

Applications of viral reverse transcriptase as a reagent to recombinant DNA problems, rather than detailed studies of its structure and mechanism, have been given the most attention in recent years. No evidence has yet been adduced for a comparable normal cellular enzyme.

49. Derse, D., Chang, Y.-C., Furman, P. A., St. Clair, M. H., and Elion, G. B. (1981) *JBC 256*, 11447.
50. Challberg, M. D., and Englund, P. T. (1979) *JBC 254*, 7812.
51. Challberg, M. D., and Englund, P. T. (1979) *JBC 254*, 7820.
52. Pedrali-Noy, G., and Spadari, S. (1979) *BBRC 88*, 1194.
53. Verma, I. (1981) *Enzymes 14*, 87; Gerard, G. F., and Grandgenett, D. P. (1980) in *Molecular Biology of RNA Tumor Viruses* (Stephenson, J. R., ed.), Academic Press, New York, Chapter 9.

## Structure

The avian enzyme, a zinc metalloprotein, is a dimer of 63- and 95-kdal subunits. In a variety of other retroviruses, the enzyme is a single polypeptide, ranging in size from 70 to 100 kdal. The variety and heterogeneity may stem from different patterns in the biosynthetic processing of the initial *gag-pol* translational product or may result from a proteolytic susceptibility that besets many other DNA polymerases (Section S6-1).

## Function

The sequence of events in transcribing the genomic RNA to a DNA duplex is becoming clearer (Section S15-7), but many questions surround the capacity of the polymerase to switch templates, to start the (+) strand on cDNA, to remove RNA, and to generate a replicative form of DNA that can integrate into the host chromosome. All reverse transcriptases have an associated RNaseH activity, but the function of an endonuclease component of the avian enzyme is still obscure.[54] Lack of DNA exonuclease activities renders replication by reverse transcriptase more error prone (especially when $Mn^{2+}$ is substituted for $Mg^{2+}$) and therefore useful for site-selective mutagenesis.[55]

Many of the extraordinary advances in molecular biology and genetic engineering have depended on the uses to which reverse transcriptase has been put.[56] These include: (i) synthesis of cDNA for cloning, sequence analysis, probes, and primers, (ii) synthesis of duplex DNA for cloning, (iii) end-labeling for sequence analysis, (iv) removal of RNA from RNA–DNA hybrids, and (v) nick translation to yield DNA of high specific radioactivity.

## S6-9 Terminal Deoxynucleotidyl Transferase[57]

Small subunits of terminal transferase (p. 226) are now recognized as proteolytic artifacts of a single polypeptide of 60 kdal found in all higher vertebrate species examined.[58] Enzymes prepared by a rapid, simple immunoadsorbent chromatography procedure from many

54. Schiff, R. D., and Grandgenett, D. P. (1980) *J. Virol. 36*, 889; Grandgenett, D. P., Golomb, M., and Vora, A. C. (1980) *J. Virol. 33*, 264.
55. Kunkel, T. A., Meyer, R. R., and Loeb, L. A. (1979) *PNAS 76*, 6331.
56. Verma, I. (1981) *Enzymes 14*, 87.
57. Ratliff, R. L. (1981) *Enzymes 14*, 105.
58. Nakamura, H., Tanabe, K. Yoshida, S., and Morita, T. (1981) *JBC 256*, 8745; Bollum, F. J., and Chang, L. M. S. (1981) *JBC 256*, 8767.

species show homologous sequences by tryptic peptide mapping analysis;[59] immunologic analysis identifies the same 60-kdal polypeptide in crude extracts from birds and humans.[60] The abundance of enzyme molecules ($10^5$) per calf thymocyte confirms earlier estimates.[61] The major conservation of size and structure of the terminal transferase, its occurrence only in certain classes of prelymphocytes, and the nature of its in vitro actions point to an important, though still obscure, role for the enzyme.[62]

An improved procedure has been described for adding homopolymer tails of each of the deoxyribonucleotides to duplex DNA with all types of 3′ termini.[63]

59. Nakamura, H., Tanabe, K., Yoshida, S., and Morita, T. (1981) *JBC 256*, 8745.
60. Bollum, F. J., and Chang, L. M. S. (1981) *JBC 256*, 8767.
61. Kung, P. C., Gottlieb, P. D., and Baltimore, D. (1976) *JBC 251*, 2399.
62. Deibel, M. R., Jr., Coleman, M. S., Acree, K. A., and Hutton, J. J. (1981) *J. Clin. Invest. 67*, 725; Yoshida, S., Masaki, S., Nakamura, H., and Morita, T. (1981) *BBA 652*, 324.
63. Deng, G., and Wu, R. (1981) *N. A. Res. 9*, 4173.

# S7

# RNA Polymerases

## Abstract

Extensive sequencing and cloning of genes and genomes have spelled out the correct size and composition of polymerase subunits; defined promoters, pause sites, attenuators, and terminators; and opened ways of observing mechanisms and controls of the many stages of transcription. However, for lack of enzymological studies, the mechanism of splicing split genes remains a grand mystery.

Binding of *E. coli* sigma subunit ($\sigma$) to two promotor regions leads to the open promoter complex that starts transcription. Sigma is also the key element of the holoenzyme that specifies which viral or plasmid genomes can be primed for DNA replication. Distinctive $\sigma$ subunits in *B. subtilis* may, by their recognition of specific promoters, explain some aspects of growth and developmental patterns, especially in sporulation and viral infections.

Rates of elongation are influenced by sequence and organization of the DNA template, and by accessory proteins (e.g., nusA) bound to the polymerase. Chain terminations either at attenuator sites or at proper termination signals also depend on accessory proteins, rho protein, for example. These transcriptional control elements augment the major regulatory influences of repressors, activating DNA-binding proteins and the increasing number of recognized posttranscriptional mechanisms.

## S7-2   Structure of *E. coli* and *B. subtilis* RNA Polymerases[1]

### *E. coli* RNA Polymerase Holoenzyme

The most reliable molecular weight for *E. coli* RNA polymerase holoenzyme, obtained from amino acid and nucleotide sequence data,[1] is 454,000. The molecular weight for the *E. coli* σ subunit[2] is 70,000, an earlier estimate of 82,000 being in error because of anomalous electrophoretic mobility in SDS gels. The DNA sequences are known for the α, σ, and β subunit genes;[3] that of β' is only partly known but is clearly unique.

### *B. subtilis* RNA Polymerase Holoenzyme

RNA polymerase holoenzyme in *B. subtilis* probably includes an additional subunit, delta (δ), a factor that enhances specificity. The possibility that δ may play a role analogous to part of the sigma subunit of Gram-negative bacteria is suggested by the sum of the sizes of *B. subtilis* σ (55 kdal) and δ (18 kdal) being near that of *E. coli* σ (70 kdal). In that event, splitting of the *B. subtilis* σ subunit gene might be related to the use of several σ factors by this organism. Among the multiple forms of *B. subtilis* RNA polymerase, the one bearing the 55-kdal σ subunit predominates ($\sigma^{55}$ RNA polymerase). Less abundant forms possess σ factors of distinct sizes ($\sigma^{28}$, $\sigma^{29}$, $\sigma^{37}$), each of which dictates recognition of promoters with distinctive sequences on the *B. subtilis* genome.[4] Control of expression of the sporulation operons is associated with $\sigma^{29}$ and $\sigma^{37}$; operons that determine growth patterns and viral (e.g., SPO1) development enlist other σ factors.[5]

1. Chamberlin, M. J. (1982) *Enzymes 15*, 61.
2. Burton, Z., Burgess, R. R., Lin, J., Moore, D., Holder, S., and Gross, C. A. (1981) *N. A. Res. 9*, 2889.
3. Chamberlin, M. J. (1982) *Enzymes 15*, 61; Squires, C., Krainer, A., Barry, G., Shen, W.-F., and Squires, C. L. (1981) *N. A. Res., 9*, 6827.
4. Haldenwang, W. G., and Losick, R. (1980) *PNAS 77*, 7000; Haldenwang, W. G., Lang, N., and Losick, R. (1981) *Cell 23*, 615; Chenchick, A., Beabealashvilli, R., and Mirzabekov, A. (1981) *FEBS Lett. 128*, 46; Gilman, M. Z., Wiggs, J. L., and Chamberlin, M. J. (1981) *N. A. Res. 9*, 5991; Moran, C. P., Jr., Lang, N., and Losick, R. (1981) *N. A. Res. 9*, 5979.
5. Losick, R., and Pero, J. (1981) *Cell 25*, 582.

S7-4 Template Binding and Site Selection
by *E. coli* RNA Polymerase

S53

SECTION S7-4:
Template Binding and Site
Selection by *E. coli*
RNA Polymerase

Promoters vary greatly in the efficiency with which they are used by RNA polymerase both in vivo and in vitro. Only relatively strong promoters form stable, tight-binding complexes with RNA polymerase in vitro. Among the parameters that influence the binding constant for RNA polymerase to a promoter, salt concentration is enormously important and can alter binding by many orders of magnitude.[6]

Binding of the σ subunit to two promoter regions leads to the open promoter complex that starts transcription. Eleven or twelve base pairs are open in the open promoter complex in which RNA polymerase can protect 60 bases in the region from −44 (upstream from the transcriptional start) to +20 (downstream).[7] The exact length of the complex and the bases involved depend on the probe used and the reaction conditions.[8] New evidence shows that T7 polymerase also opens base pairs at promoter regions.[9] Sigma interacts directly or sequentially with DNA regions[10] centered at −35 and −10 upstream from the transcriptional start. In the open promoter complex, σ can be crosslinked to the bases in the region between −10 and the start site.[11] Free *E. coli* σ factor can bind to DNA[12] as does the σ factor of *B. subtilis*.[13] Conserved promoter sequences are characteristic of the recognition specificity of bacterial RNA polymerase;[14] in *B. subtilis*, different promoters are used by different viral and bacterial σ factors.[15] When the bacterial polymerase contains an altered σ factor, altered DNA base sequences are found at two regions of the promoter.

6. de Haseth, P. L., Lohman, T. M., Burgess, R. R., and Record, M. T., Jr. (1978) *B. 17*, 1612; Strauss, H. S., Burgess, R. R., and Record, M. T., Jr. (1980) *B. 19*, 3496, 3504; Chamberlin, M., Kadesch, T., and Rosenberg, S. (1982) in *Promoters: Structure and Function* (Rodriguez, R., and Chamberlin, M., eds.) Praeger Press, New York.

7. Siebenlist, U., Simpson, R. B., and Gilbert, W. (1980) *Cell 20*, 269.

8. Moran, C. P., Jr., Lang, N., and Losick, R. (1981) *N. A. Res. 9*, 5979.

9. Strothcamp, R. E., Oakley, J. L., and Coleman, J. E. (1980) *B. 19*, 1074.

10. Haldenwang, W. G., and Losick, R. (1980) *PNAS 77*, 7000; Haldenwang, W. G., Lang, N., and Losick, R. (1981) *Cell 23*, 615; Chenchick A., Beabealashvilli, R., and Mirzabekov, A. (1981) *FEBS Lett. 128*, 46.

11. deHaseth, P. L., Lohman, T. M., Burgess, R. R., and Record, M. T., Jr. (1978) *B. 17*, 1612; Strauss, H. S., Burgess, R. R., and Record, M. T., Jr. (1980) *B. 19*, 3496, 3504; Chamberlin, M., Kadesch, T., and Rosenberg, S. (1982) in *Promoters: Structure and Function* (Rodriguez, R., and Chamberlin, M., eds.) Praeger Press, New York.

12. Stender, W. (1980) *N. A. Res. 8*, 3459; Kudo, T., and Doi, R. H. (1981) *JBC 256*, 9778.

13. Kudo, T., Jaffe, D., and Doi, R. H. (1981) *MGG 181*, 63.

14. Wiggs, J. L., Gilman, M. Z., and Chamberlin, M. J. (1981) *PNAS 78*, 2762.

15. Haldenwang, W. G., and Losick, R. (1980) *PNAS 77*, 7000; Haldenwang, W. G., Lang, N., and Losick, R. (1981) *Cell 23*, 615; Chenchick, A., Beabealashvilli, R., and Mirzabekov, A. (1981) *FEBS Lett. 128*, 46; Gilman, M. Z., Wiggs, J. L., and Chamberlin, M. J. (1981) *N. A. Res., 9*, 5991; Lee, G., and Pero, J. (1981) *JMB 152*, 247.

RNA polymerase likely locates a promoter by a "sliding" mechanism in which it moves rapidly along DNA in a one-dimensional diffusion process, being released about every 1000 base pairs. This scheme follows from the very rapid rates at which RNA polymerase locates promoter sites, rates far in excess of those possible by free diffusion.[16]

Nonspecific complexes of *E. coli* RNA polymerase holoenzyme with DNA vary enormously in structure and stability[17] from those that are very weak to complexes that approach promoter-site specificity. They differ in thermodynamic and kinetic binding constants and in their responses to salt and temperature.

RNA polymerase is also employed in *E. coli* in some instances to initiate and regulate DNA replication. Its exact role in the start of an *E. coli* or phage λ chromosome replication cycle is still undefined, but replication of certain plasmids, such as ColE1, depends on an RNA transcript that is covalently extended by DNA polymerase I to start the DNA chain (Section S14-9). Replication of the single-stranded filamentous phages (e.g., M13) is also primed by RNA polymerase, whereas replication of the single-stranded polyhedral phages (e.g., φX174) is not. This remarkable discrimination between M13 and φX174 depends on the σ factor in the *E. coli* holoenzyme and provides an easy and decisive assay for an active form of this subunit (Section S11-8).

Transcription also serves in regulating the start of replication of some plasmids. In the case of ColE1 and IncFII (R1, R100, R6-5) plasmids, an RNA transcript from a region upstream from the origin inhibits replication and so provides a mechanism for controlling the number of plasmids per cell (Section S14-9).

## S7-5 Chain Initiation and Elongation by *E. coli* RNA Polymerase

Initiation with UTP and CTP can occur at certain promoters in vivo and in vitro although it is less common than with ATP and GTP. The rate of initiation may be substantial even at 10 $\mu$M ATP for some promoters, but increases up to 1 mM.[18] For certain strong promoters (e.g., A1 or A2 of phage T7), chain initiation is rapid and essentially irreversible. For the *lacUV5* promoter, by contrast, the enzyme in vitro

16. von Hippel, P., Bear, D., Winter, R., and Berg, O. (1982) in *Promoters: Structure and Function* (Rodriguez, R., and Chamberlin, M., eds.) Praeger Press; New York; Chamberlin, M., Kadesch, T., and Rosenberg, S. (1982) ibid; Bujard, H., Niemann, A., Breunig, K., Roisch, U., Dresel, A., von Gabain, A., Gentz, R., Stüber, D., and Weiher, H. (1982) ibid.
17. deHaseth, P. L., Lohman, T. M., Burgess, R. R., and Record, M. T., Jr. (1978) *B. 17.* 1612; Lohman, T. M., Wensley, C. G., Cina, J., Burgess, R. R., and Record, M. T., Jr. (1980) *B. 19,* 3516; Kadesch, T. R., Williams, R. C., and Chamberlin, M. J. (1980) *JMB 136,* 65, 79.
18. Nierman, W. C., and Chamberlin, M. J. (1979) *JBC 254,* 7921.

has difficulty starting a long chain and goes through many rounds of abortive syntheses, releasing di-, tri-, and oligonucleotides.[19] This recycling is favored at low rNTP concentrations and is less significant as rNTP concentrations approach in vivo levels. The σ subunit is probably not released from the complex until about eight nucleotides are present in the nascent RNA chain.[20]

The elongation rate in vitro is 17 to 30 nucleotides per second for core polymerase and is even slower for the enzyme with nusA protein bound to it[21] (Section S7-6). With the slower (nusA-bound) in vitro rate only one-tenth of that in vivo, it seems likely that additional factors and possibly an energy input are involved in elongation.[22] Furthermore, since the DNA template in vivo has a complex chromatin-like structure, its protein components may play a role in the rate and specificity of transcription.[23]

Pausing by an RNA polymerase molecule during the course of transcription may have an important regulatory function. In the *trp* operon, a pause after synthesis of ninety nucleotides into the leader region (Section S7-6) may help synchronize transcription and translation in a way that enables the downstream attenuator to exercise its control.[24] A set of pause sites in the ribosomal RNA operon, responsive to the regulatory factors ppGpp and nusA protein, may also contribute to the control of rDNA expression.[25]

# S7-6    Chain Termination by
## *E. coli* RNA Polymerase

Chain termination not only is as much a part of transcription as initiation but can also provide a powerful regulatory brake on initiation events.[26] By the process of *attenuation,* transcription of some operons is turned off before the site of the structural gene is reached. This is true for some amino acid biosynthesis operons when the particular amino acid is abundant. Concurrent translation of a leader region (the segment of the operon between the transcription start site and the structural gene) produces a short peptide enriched with a tandem array of the amino acid controlled by the operon. When the

---

19. Carpousis, A. J., and Gralla, J. D. (1980) *B. 19,* 3245.
20. Hansen, U. M., and McClure, W. R. (1980) *JBC 255,* 9556.
21. Kassavetis, G. A., and Chamberlin, M. J. (1981) *JBC 256,* 2777; Kingston, R. E., and Chamberlin, M. J. (1981) *Cell 27,* 523.
22. Kassavetis, G. A., and Chamberlin, M. J. (1981) *JBC 256,* 2777; Greenblatt, J. (1981) *Cell 24,* 8; Greenblatt, J., and Li, J. (1981) *Cell 24,* 421.
23. Lathe, R., Buc, H., Lecocq, J.-P., and Bautz, E. K. F. (1980) *PNAS 77,* 3548.
24. Winkler, M. E., and Yanofsky, C. (1981) *B. 20,* 3738.
25. Kingston, R. E., and Chamberlin, M. J. (1981) *Cell 27,* 523.
26. Yanofsky, C. (1981) *Nat. 289,* 751.

amino acid is in short supply, ribosomes and translation are stalled at one of the codons for that amino acid. Due to this lapse in translation, the transcript is free to form a particular secondary structure (stem and loop) that allows the transcribing RNA polymerase molecule to proceed through the attenuator, the template site that otherwise specifies a termination of transcription. In this kinetically governed event, the formation of secondary structure in a stalled leader-translation complex directly prevents formation of an alternative secondary structure of the attenuator that is the signal for termination.

Rho protein (rho)[27] has an important role in some but probably not all termination events. Rho is responsible for the release of the polymerase at certain pause sites[28] but not at others. Some factor other than rho appears to be involved in RNA polymerase release at the trp attenuator site[29] and at the termination of the phage T7 early transcript.[30] Rho may complex with RNA polymerase, at least transiently, inasmuch as certain RNA polymerase mutants are known to be altered in terminations dependent on rho.[31]

The nature of the terminating influence of a DNA region distal to the termination sequence for the trp operon (p. 249) has now been shown to be due to a second termination signal, rather than a direct involvement of this region in the functioning of the primary termination signal.[32]

The role of N protein (encoded by phage λ)[33] in antitermination depends on two proteins, nusA and nusB, at least one of which, nusA, appears to be an RNA polymerase "elongation factor." The nusA protein dramatically affects the elongation of RNA chains in vitro and particularly RNA polymerase interaction with pause sites, attenuators, and terminators.[34] Binding studies with purified nusA and core polymerase suggest competition with σ factor and the existence of alternative forms of the holoenzyme, one with σ for promoter recognition and another with nusA for elongation. In binding the nusA form, factors such as N protein of phage λ may exert their potent regulatory influence.[35]

27. Finger, L. R., and Richardson, J. P. (1981) B. 20, 1640.
28. Kassavetis, G. A., and Chamberlin, M. J. (1981) JBC 256, 2777.
29. Yanofsky, C. (1981) Nat. 289, 751.
30. Kiefer, M., Neff, N., and Chamberlin, M. J. (1977) J. Virol. 22, 548.
31. Guarente, L. (1979) JMB 129, 295.
32. Platt, T. (1981) Cell 24, 10.
33. Greenblatt, J., and Li, J. (1982) JBC 257, 362.
34. Kingston, R. E., and Chamberlin, M. J. (1981) Cell 27, 523; Greenblatt, J. (1981) Cell 24, 8; Greenblatt, J., and Li, J. (1981) Cell 24, 421.
35. Kung, H.-F., and Weissbach, H. (1980) Arch. Biochem. Biophys. 201, 544; Greenblatt, J., and Li, J. (1981) JMB 147, 11; Greenblatt, J. and Li, J. (1981) Cell 24, 421.

## Initiation and Elongation

Isolated eukaryotic RNA polymerases initiate chains on DNA templates randomly, primarily at ends, breaks, or gaps, rather than at specific promoter sequences. This lack of specificity can be overcome by addition of nuclear protein fractions, which program the polymerase to use specific promoters.[36] For the eukaryotic RNA polymerase III, one such factor selectively directs synthesis of 5S RNA transcripts. This factor is a DNA-binding protein that interacts specifically with sequences well within the gene for the 5S RNA gene.[37]

Activation of promoters for eukaryotic RNA polymerase II is even more complex and depends on their location in the chromosome. For example, expression of the provirus of mouse mammary tumors (MMTV), with its strong, hormonally induced promoter, is influenced by the activity of neighboring DNA sequences.[38] "Activation" of a DNA region for transcription may involve changes in chromatin structure that can be monitored by accessibility to E. coli RNA polymerase[39] and to nucleases. In SV40 DNA, the 72-bp repeat that starts 115 bp upstream from the early mRNA start sites is crucial for expression of the early genes.[40] Remarkably, the cis-acting effect of this sequence is independent of its orientation and can operate over very long distances, suggestive of a bidirectional entry site for RNA polymerase. Similar "enhancer" sequences are present in region A of polyoma DNA but lack a strong homology with the 72-bp sequence of SV40.[41]

As with prokaryotic RNA polymerases, the rate of chain elongation by RNA polymerase II in vitro is only about one-tenth the in vivo rate. The slower rate may be due to transcriptional pause sites distributed along the DNA template that exaggerate the normal delays. Possibly, too, alterations in the isolated polymerase or the absence

35a. Lewis, M. K., and Burgess, R. R. (1982) *Enzymes 15*, 110.
36. Weil, P. A., Luse, D. S., Segall, J., and Roeder, R. G. (1979) *Cell 18*, 469; Weil, A., Segall, J., Harris, B., Ng, S.-Y., and Roeder, R. G. (1979) *JBC 254*, 6163.
37. Engelke, D. R., Ng, S.-Y., Shastry, B. S., and Roeder, R. G. (1980) *Cell 19*, 717; Sakonju, S., Brown, D. D., Engelke, D. R., Ng, S.-Y., Shastry, B. S., and Roeder, R. G. (1981) *Cell 23*, 665.
38. Feinstein, S., Ross, S., and Yamamoto, K. (1982) *JMB*, submitted.
39. Craine, B. L., and Kornberg, T. (1981) *Cell 25*, 671.
40. Benoist, C., and Chambon, P. (1981) *Nat. 290*, 304; Gruss, P., Dhar, R., and Khoury, G. (1981) *PNAS 78*, 943; Moreau, P., Hen, R., Wasylyk, B., Everett, R., Gaub, M. P., and Chambon, P. (1981) *N. A. Res. 9*, 6047; Banerji, J., Rusconi, S., and Schaffner, W. (1981) *Cell 27*, 299; Fromm, M., and Berg, P., personal communication.
41. Tyndall, C., La Mantia, G., Thacker, C. M., Favaloro, J., and Kamen, R. (1981) *N. A. Res. 9*, 6231; de Villiers, J., and Schaffner, W. (1981) *N. A. Res. 9*, 6251.

of accessory factors accounts for the slow rate and for anomalously long DNA–RNA hybrids.[42]

## Mitochondrial RNA Polymerase

The polymerase isolated from yeast mitochondria[43] ranges in size from 100 to 150 kdal and is made up of 45-kdal subunits. Differing from nuclear and bacterial RNA polymerases in size, it also has distinctive enzymatic properties. Inhibition by $Mn^{2+}$, salt sensitivity, and a preference for a poly d(AT) over a natural template distinguish the mitochondrial enzyme from the nuclear one. The preference for poly d(AT) may reflect the 82-percent AT content of the mitochondrial genome. The mitochondrial polymerase differs from bacterial polymerases in its resistance to rifampicin and streptolydigin and its $Mn^{2+}$ and salt sensitivity. It also differs from many phage-encoded polymerases (T7, N4; see below), which are single subunit enzymes with fastidious specificities for their homologous templates.[44]

## Splicing

Processing of transcripts to remove the numerous regions (introns) that interrupt the coding sequences (exons) making up the structural gene is a remarkable and distinctive feature of eukaryotic gene expression. Transcripts with as many as 33 introns, introns several thousand base pairs long, and genes 20 times longer than the corresponding messenger RNA have been described.[45] Aside from the special instance of removing one intron from yeast tRNA (p. 276), the mechanism of splicing split genes remains a mystery.

## S7-8  RNA Polymerase in Viral Infections

### Phage T7

The gene for T7 RNA polymerase encodes a polypeptide of 98 kdal.[46] Several minor T7 promoters for the T7 phage polymerase, essential for transcription of T7 class II genes, are located in the "early"

42. Dezélée, S., Sentenac, A., and Fromageot, P. (1974) *JBC 249*, 5978; Lescure, B., Chestier, A., and Yaniv, M. (1978) *JMB 124*, 73; Kadesch, T. R., and Chamberlin, M. J., personal communication.
43. Levens, D., Morimoto, R., and Rabinowitz, M. (1981) *JBC 256*, 1466; Levens, D., Lustig, A., and Rabinowitz, M. (1981) *JBC 256*, 1474.
44. Chamberlin, M., and Ryan, T. (1982) *Enzymes 15*, 87.
45. Chambon, P. (1981) *Sci. Amer. 244* (no. 5), 60; Breathnach, R., and Chambon, P. (1981) *ARB 50*, 349.
46. Stahl, S. J., and Zinn, K. (1981) *JMB 148*, 481.

where previously thought. Some promoter sequences (p. 256) are now known to be in error, and correct sequences are available for the early T7 region.[48]

## Phages T3, PBS2, SPO1, N4

Although homologies between T3 and T7 promoters exist, their use by the heterologous enzyme[49] is relatively ineffective. The question of the early PBS2 transcription mechanism has grown more complex.[50] The SPO1 polypeptides have now been shown clearly to act as σ replacement factors.[51] The N4 RNA polymerase, remarkable for requiring the action of an ATPase for transcription, includes that activity within its 350-kdal polypeptide.[52] In addition, the *E. coli* single-strand binding protein is essential for transcription of the denatured N4 DNA template.[53]

## Vaccinia

For in situ transcription of the vaccinia nucleoprotein template, the virion RNA polymerase depends, as does the phage N4 polymerase, on the simultaneous action of an ATPase.[54] This is not true for the purified vaccinia polymerase acting on naked DNA. Such behavior resembles that of *E. coli* DNA polymerase III holoenzyme, which depends on a helicase and ATP consumption (Section S9-10) to expose the template of a replicating rolling circle.

47. Carter, A. D., Morris, C. E., and McAllister, W. T. (1981) *J. Virol. 37*, 636; Chamberlin, M., and Ryan, T. (1982) *Enzymes 15*, 87.
48. Dunn, J. J., and Studier, W. F. (1981) *JMB 148*, 303.
49. Chamberlin, M., and Ryan, T. (1982) *Enzymes 15*, 87.
50. Chamberlin, M., and Ryan, T. (1982) *Enzymes 15*, 87.
51. Wiggs, J. L., Gilman, M. Z., and Chamberlin, M. J. (1981) *PNAS 78*, 2762.
52. Rothman-Denes, L., personal communication.
53. Chamberlin, M., and Ryan, T. (1982) *Enzymes 15*, 87.
54. Shuman, S., Spencer, E., Furneaux, H., and Hurwitz, J. (1980) *JBC 255*, 5396.

# S8

# Ligases

## Abstract

The DNA ligase gene of phage T4, having been cloned and amplified, can furnish the amounts of enzyme needed to obtain the long-needed information about its structure. The primary structure of T7 DNA ligase is now known from its DNA sequence. The requirement for a DNA ligase in yeast is established with identification of a cell-development-cycle mutant that lacks it. The RNA ligase gene of phage T4 has a domain to encode its nucleic acid joining activity separate from that responsible for tail-fiber attachment. A linkage between RNA ligase and polynucleotide kinase, seen in the pheno-types of their mutants, is the latest clue to their physiological functions.

## S8-2   Properties, Abundance, and Amplification[1]

A 1.9-kb fragment of T4 DNA containing the DNA ligase gene has been inserted into a λ cloning vector.[1a] An E. coli lysogen containing this prophage can be induced to synthesize large amounts of en-

---

1. Engler, M. J., and Richardson, C. C. (1982) Enzymes 15, 3.
1a. Wilson, G. G., and Murray, N. E. (1979) JMB 132, 471.

zyme.[2] Several purification procedures have been described,[3] which should provide sufficient protein to permit detailed studies of the physical properties of this enzyme.

The complete amino acid sequence of T7 DNA ligase can be deduced from the DNA sequence.[4] Its molecular weight is 41,133.

## S8-6   Ligases of Eukaryotes

A conditionally lethal mutant of the yeast *Schizosaccharomyces pombe* has been identified as a DNA ligase mutation.[5] At the permissive temperature (25°C) it has less than 20 percent of the normal activity, while at the restrictive temperature (35°C) no activity can be detected. At the restrictive temperature, mutant cells can replicate DNA during S phase, but mitosis does not occur. Most of the new DNA that is synthesized at the restrictive temperature remains as very small pieces, suggesting that DNA ligase is needed for joining of the fragments. The mutant cells are also highly sensitive to ultraviolet light at the nonpermissive temperature. A similar mutant of *Saccharomyces cerevisiae* shows similar properties but has no detectable DNA ligase activity at either the permissive or nonpermissive temperature.[6]

## S8-8   RNA Ligase[7]

The tail-fiber attachment activity of phage T4 gene 63 protein appears to be completely independent of its RNA ligase activity. Mutants (*rli⁻*) have been isolated that lack the RNA ligase activity but are normal in tail-fiber attachment.[7a] Such mutants grow normally in wild-type *E. coli* but are defective in T4 DNA replication and particularly in T4 late gene expression in certain restrictive *E. coli* strains. This phenotype is identical to that of *pseT* mutations in the T4 polynucleotide kinase gene (Section S8-9), suggesting that the ligase and kinase take part in the same (still unknown) pathway.

An unusual RNA ligase activity in wheat germ extracts[8] forms an alkali- and RNase-stable linkage during conversion of a linear polyribonucleotide to a covalently closed circle. The linkage between the

---

2. Murray, N. E., Bruce, S. A., and Murray, K. (1979) *JMB 132*, 493.
3. Murray, N. E., Bruce, S. A., and Murray, K. (1979) *JMB 132*, 493; Tait, R. C., Rodriguez, R. L., and West, R. W., Jr. (1980) *JBC 255*, 813; Davis, R. W., Botstein, D., and Roth, J. R. (1980) in *Advanced Bacterial Genetics*, CSHL, p. 196.
4. Dunn, J. J., and Studier, F. W. (1981) *JMB 148*, 303.
5. Nasmyth, K. A. (1977) *Cell 12*, 1109.
6. Johnston, L. H., and Nasmyth, K. A. (1978) *Nat. 274*, 891.
7. Uhlenbeck, O. C., and Gumport, R. I. (1982) *Enzymes 15*, 31.
7a. Runnels, J. M., Soltis, D., Hey, T., and Snyder, L. (1982) *JMB 154*, 273.
8. Konarska, M., Filipowicz, W., Domdey, H., and Gross, H. J. (1981) *Nat. 293*, 112; Konarska, M., Filipowicz, W., and Gross, H. J. (1982) *PNAS 79*, 1474.

termini contains a 2'-phosphomonoester as well as the standard 3',5'-phosphodiester bridge. The reaction requires a 5'-hydroxyl terminus and a 2',3'-cyclic phosphate terminus as well as ATP. The mechanism of this joining reaction and what potential this ester may have for branching of the RNA remain unknown. The relation of this wheat germ activity to the ligase activities in yeast and *Xenopus* oocytes[9] involved in splicing the introns from precursor tRNAs is also unclear.

## S8-9   Polynucleotide Kinases

Phage polynucleotide kinase is a novel enzyme and vital biochemical reagent. It is accorded this separate section to elevate it from a list of curious phage-induced proteins (p. 535). Furthermore, the properties of such kinases and their wide distribution in animal cells argue for a possible role linked to DNA ligase in DNA repair reactions.

T4 Phage Polynucleotide Kinase[10]

The enzyme is induced in *E. coli* by the T-even phages but has not been detected in uninfected bacteria. Purified to homogeneity from phage T4-infected cells, it is a dimer or tetramer of identical 34-kdal polypeptides.[11] The polypeptide is a product of the *pseT* gene located between genes 63 and 31 on the phage T4 map (Fig. S14-3).[12]

The enzyme uses virtually all nucleoside triphosphates to phosphorylate the 5'-hydroxyl terminus of DNA, RNA, and oligonucleotides; even nucleoside 3'-monophosphates and terminal nucleotides with bases protected by bulky adducts are excellent substrates. However, nicked DNA is a very poor acceptor; protruding ends of duplexes are preferred over flush or recessed ends. Unlike most kinases, this enzyme catalyzes two other reactions: (i) ready reversal of the phosphorylation reaction and thereby an exchange between the γ-phosphate of an NTP donor and the 5'-phosphoryl group of a polynucleotide, and (ii) phosphatase action on 3'-phosphoryl groups of DNA and oligodeoxynucleotides. Mutations in *pseT* generally affect both the kinase and phosphatase activities, but one mutant (*pseT1*) produces an altered enzyme deficient only in phosphatase.[13]

The enzyme is an indispensable reagent for labeling DNA and

9. Nishikura, K., and de Robertis, E. M. (1981) *JMB 145*, 405.
10. Richardson, C. C. (1981) *Enzymes 14*, 299; Kleppe, K., and Lillehaug, J. R. (1979) *Adv. Enz. 48*, 245.
11. Lillehaug, J. R. (1977) *EJB 73*, 499; Soltis, D., and Uhlenbeck, O. C., personal communication.
12. Sirotkin, K., Cooley, W., Runnels, J., and Snyder, L. R. (1978) *JMB 123*, 221; Mileham, A. J., Revel, H. R., and Murray, N. E. (1980) *MGG 179*, 227.
13. Cameron, V., Soltis, D., and Uhlenbeck, O. C. (1978) *N. A. Res. 5*, 825.

RNA chains for sequence analysis, monitoring the progress of DNA and RNA synthesis, identifying the ends of chains and determining their lengths, fingerprinting, physical mapping of restriction enzyme fragments, and still other uses. However, the function of the enzyme in phage development is not clear. The combination of 5'-kinase and 3'-phosphatase activities in one oligomeric enzyme suggests a role in preparing DNA breaks with 3' phosphates for covalent closure by DNA ligases. Yet, phenotypic defects in DNA repair, recombination, or replication have not been demonstrated with *pseT* mutants in wild-type cells. However, the kinase shares a function with the phage RNA ligase in supporting T4 late gene expression in certain host strains (Section S8-8).

## Eukaryotic DNA Kinase[14]

Unlike the phage-encoded enzyme, this polynucleotide kinase is specific for DNA, works at nicks in duplex DNA, and requires an oligonucleotide at least ten residues long (Table S8-1). Used in conjunction with the phage enzyme, RNA and DNA chain ends can be distinguished, as can nicks and free ends. DNA kinase purified from nuclei of rat liver or calf thymus is a polypeptide of 80 kdal. When freed of an RNA kinase, which likely functions in mRNA capping,[15] the strict specificity for DNA becomes apparent.[16]

---

14. Zimmerman, S. B., and Pheiffer, B. H. (1981) *Enzymes 14*, 315; Kleppe, K., and Lillehaug, J. R. (1979) *Adv. Enz. 48*, 245.
15. Shuman, S., and Hurwitz, J. (1979) *JBC 254*, 10396.
16. Tamura, S., Teraoka, H., and Tsukada, K. (1981) *EJB 115*, 449.

TABLE S8-1
Properties of polynucleotide, DNA, and RNA kinases

| | Phage T4 polynucleotide kinase | Eukaryote | |
|---|---|---|---|
| | | DNA kinase | RNA kinase |
| Mass, kdal | 68 | 80 | |
| pH optimum | 7.7 | 5.5 | 8.4 |
| $SO_4^{2-}$ sensitivity | low | high | low |
| Acceptor specificity | | | |
|   RNA vs. DNA | RNA and DNA | DNA | RNA |
|   Nicks in DNA | no | yes | |
|   Size | mono to poly | oligo ($>$10) to poly | poly |
| Donor NTP | | | |
|   Specificity | broad | broad | |
|   ATP, $K_m$ | 14 | 2 | 500 |
| 3'-Phosphatase activity | yes | | |
| Exchange activity | yes | yes | |

# S9

# Binding and Unwinding
# Proteins and Topoisomerases

## Abstract

In a new edition of *DNA Replication,* this single chapter would
need to be expanded and divided into four in order to present the
current knowledge and significance of: (i) the duplex DNA-binding
proteins that organize genomes and regulate their expression, (ii) the
single-strand binding proteins, essential for replication, repair, and
recombination, (iii) the energy-driven helicases and related DNA-
dependent ATPases, and (iv) the remarkable topoisomerases that
control DNA conformation.

Crystallographic analysis and solution studies of repressor-operator
interactions show how the protein domains interact with the DNA
grooves and with other factors to achieve their physiological effects.
The numerous single-strand binding proteins go on and off DNA in
treadmill fashion (resembling tubulin polymerization) to protect and

conform it for replication, repair, and recombination; the eukaryotic proteins generally differ from the prokaryotic in not displaying highly cooperative binding and being modulated by phosphorylation. How histone-like proteins organize the *E. coli* nucleoid and how the eukaryotic histones and related proteins achieve the higher order folding of nucleosomes in chromatin are becoming better understood.

The helicases use ATP energy to move processively and unidirectionally along one or the other strand, opening the duplex in advance of replication. These and the numerous other DNA-dependent ATPases with diverse functions appear to operate on a "buy now, pay later" plan: Conformational changes induced by ATP achieve the desired effect, and ATP hydrolysis restores the initial state for a cyclical sequence.

The many topoisomerases operate on a common principle of concerted breakage and reunion of DNA strands to make the duplex more or less supercoiled; type I topoisomerases break one strand and type II both strands. Reversible catenation and knotting of DNA rings are among the consequences of DNA strand passage by topoisomerases. The covalent complex between a tyrosine side chain and a phosphoryl moiety at the break conserves the energy of the phosphodiester bond for subsequent resealing.

## S9-1 DNA-Binding Proteins[1]

Proteins that bind DNA elicit and reflect its dynamic structural and functional features. Among the best understood in their three-dimensional structure and interactions with DNA are the repressors and activators that regulate gene expression.

### Cro Protein

Analysis of the crystal structure of the phage λ cro repressor (cro) suggests a gratifyingly simple and symmetrical way in which it may bind DNA.[2] Cro is a small protein that together with the cI protein (λ repressor) determines whether the phage will follow a lysogenic or lytic pathway.[3] The structure of the monomer, containing only 66 amino acids (mol. wt. 7351), is known at 2.8-Å resolution. It consists of three extended segments that form an antiparallel β sheet and three α helices. An extended carboxy-terminal tail interacts with

---

1. Kowalczykowski, S. C., Bear, D. G., and von Hippel, P. H. (1981) *Enzymes 14*, 373; Geider, K., and Hoffmann-Berling, H. (1981) *ARB 50*, 233.
2. Anderson, W. F., Ohlendorf, D. H., Takeda, Y., and Matthews, B. W. (1981) *Nat. 290*, 754.
3. Folkmanis, A., Takeda, Y., Simuth, J., Gussin, G., and Echols, H. (1976) *PNAS 73*, 2249.

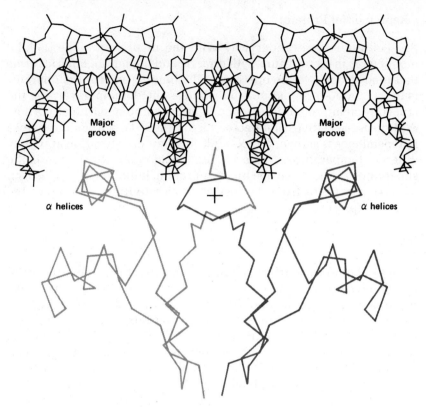

**FIGURE S9-1**
Presumed interaction of cro repressor with DNA. Two cro monomers, related by twofold symmetry axis (+ symbol), interact with DNA such that a pair of twofold-related α helices occupy successive major grooves of the DNA and two extended polypeptide strands run parallel to the backbone of the DNA. (Courtesy of Professor B. W. Matthews.)

that of another monomer to form a dimer with twofold rotational symmetry. Similarly, the three 17-base-pair operators in the promoter region of λ DNA each have palindromic, or twofold rotational, symmetry.[4] This leads to a proposal[5] for the association of the cro dimer with operator DNA in the standard B-form in which a pair of symmetrically related α helices from the dimer bind to the two halves of the palindromic sequence found within successive major grooves of the DNA (Figure S9-1).

4. Ptashne, M., Jeffrey, A., Johnson, A. D., Maurer, R., Meyer, B. J., Pabo, C. O., Roberts, T. M., and Sauer, R. T. (1980) *Cell 19*, 1.
5. Anderson, W. F., Ohlendorf, D. H., Takeda, Y., and Matthews, B. W. (1981) *Nat. 290*, 754.

λ Repressor (cI Protein)

Parallel studies are under way with the cI protein.[6] The 26-kdal monomer is in equilibrium with dimeric and higher oligomeric forms that bind strongly to the operators. The binding site for the operator is in the amino-terminal portion whereas the domain governing the crucial protein-protein interaction is in the carboxy-terminal portion. A protease-sensitive "connector" joins these two domains whose independence is shown, for example, by their denaturation at widely different temperatures.[7] This 40-amino-acid connector region is the site for proteolytic cleavage by recA protein; induction of a λ lysogen (p. 634) may result from the loss of cooperativity in binding of the cleaved repressor.

CAP

Structural studies of the catabolite gene activator protein (CAP) of E. coli have progressed to 2.9-Å resolution.[8] The active form of CAP is a dimer of identical subunits of 201 amino acids. Each subunit is folded into distinct amino-terminal and carboxy-terminal domains connected by a 4- to 5-amino-acid hinge. The morphology revealed in the subunit and dimer offer opportunities for speculation about the binding of cyclic AMP and DNA that suggest mechanisms[9] for promoting polymerase transcription of catabolite-sensitive genes.

Domains

A recurring theme in these studies of the cro and λ repressors, the CAP activator, the gene 32 protein (Section S9-3), the gene 5 protein (Section S9-4), and the lac repressor is the importance of domains in the organization and function of these DNA-binding proteins: domains to recognize DNA, domains to promote oligomerization, domains for interaction with other proteins, and functional domains connected by hinge regions susceptible to proteolytic cleavage.[10]

6. Ptashne, M., Jeffrey, A., Johnson, A. D., Maurer, R., Meyer, B. J., Pabo, C. O., Roberts, T. M., and Sauer, R. T. (1980) Cell 19, 1.
7. Pabo, C. O., Sauer, R. T., Sturtevant, J. M., and Ptashne, M. (1979) PNAS 76, 1608.
8. McKay, D. B., and Steitz, T. A. (1981) Nat. 290, 744.
9. Kolb, A., and Buc, H. (1982) N. A. Res. 10, 473.
10. Hosoda, J., Burke, R. L., Moise, H., Kubota, I., and Tsugita, A. (1980) ICN–UCLA Symposia 19, 507; Williams, K. R., Sillerud, L. O., Schafer, D. E., and Konigsberg, W. H. (1979) JBC 254, 6426; Burke, R. L., Alberts, B. M., and Hosoda, J. (1980) JBC 255, 11484; Ogata, R. T., and Gilbert, W. (1978) PNAS 75, 5851; von Hippel, P. H., Bear, D. G., Winter, R. B., and Berg, O. G. (1981) in Promoters: Structure and Function (Rodriguez, R., and Chamberlin, M., eds.), Praeger Press, New York.

The DNA-binding domain of the lac repressor proves to be homologous with the sequences of five DNA-binding proteins, including the cro and λ repressors.[11]

## DNA Site Selection[12]

It is unlikely that a site-specific binding protein finds its cognate DNA sequence by tediously testing all possible positions along the DNA. A three-stage interaction could be involved in which binding becomes increasingly specific. Stage I: Territorial binding results from the interaction of cationic amino acids with the negatively charged cloud surrounding DNA.[13] This limits the protein to its substrate, where it moves freely and rapidly along the length of the DNA without expenditure of energy. Stage II: The rapid scan is slowed when the protein interacts preferentially with some conformational feature of DNA, such as supercoiling, single-strandedness, or backbone structure. AT richness, GC richness, and inverted repeats of DNA sequences may form the basis of these conformational cues for binding. Stage III: With region-specific binding achieved, fewer interactions need to be tried to achieve site specificity. Direct interactions in the major or minor grooves may suffice or binding to groups on a single strand may follow localized melting of the duplex.

## RNA-Binding Proteins

An exception to the arbitrary exclusion of RNA subjects from *DNA Replication* was made for RNA polymerases (Chapter 7) and should also be made for RNA-binding proteins. Important principles can be learned about the proteins that form discrete particles[14] in addition to those that constitute ribosomes and viruses. One such protein from *Xenopus* binds both RNA and DNA.[15] It associates with 5S RNA to form a stable 7S particle that accumulates in young oocytes in anticipation of accelerated ribosome synthesis. The protein is also an essential positive transcription factor that binds specifically to the center of the gene encoding the 5S RNA.

11. Matthews, B. W., Ohlendorf, D. H., Anderson, W. F., and Takeda, Y. (1982) *PNAS 79*, 1428.
12. Berg, O. G., Winter, R. B., and von Hippel, P. H. (1981) *B. 20*, 6929.
13. Manning, G. S. (1978) *Q. Rev. Biophys. 11*, 179; (1980) *Biopolymers 19*, 37; Wilson, R. W., and Bloomfield, V. A. (1979) *B. 18*, 2192.
14. Picard, B., and Wegnez, M. (1979) *PNAS 76*, 241; Lerner, M. R., and Steitz, J. A. (1981) *Cell 25*, 298.
15. Pelham, H. R. B., and Brown, D. D. (1980) *PNAS 77*, 4170; Sakonju, S., Brown, D. D., Engelke, D., Ng, S.-Y., Shastry, B. S., and Roeder, R. G. (1981) *Cell 23*, 665.

## S9-2    Single-Strand Binding Proteins[16]

The remarkable single-strand binding proteins, without apparent enzymatic activity, perform essential functions in replication, repair, and recombination, and as may be inferred from in vivo studies, can also exert a less direct control over these processes. Several generalizations about their properties (pp. 278–282) deserve further emphasis.

(i)    They act stoichiometrically but can be cycled in a treadmill fashion, going on and off the ends of an aggregate, as tubulin does in solution.[17] The limited supplies of the E. coli and T4 phage proteins (Sections S9-3, S9-5) must be used in this "catalytic" fashion, as opposed to the relatively stoichiometric use of the large depots of M13 phage and adenoviral proteins (Sections S9-4, S9-6) that coat entire genomes.

(ii)    The properties of one kind of single-strand binding protein (e.g., E. coli SSB) distinguish it strikingly from those of others that populate the same cell (e.g., T4, M13, T7 SSBs). These unique features are manifested both in how the DNA is bound and in interactions with other proteins.

(iii)    Dissociation of the protein from DNA can be as important in its overall function as binding to the DNA and may rely even more on the auxiliary actions of other proteins and competitive DNAs.

(iv)    Although cellular levels of the binding proteins are regulated to achieve optimal function, deviations can be tolerated and may perhaps even be induced.

(v)    Helix destabilization is only one feature of these binding proteins and not so reproducible at that.[18] Hence, the more general term of single-strand binding protein (SSB) is gaining acceptance for the E. coli protein, especially to conform to the term ssb for its genetic locus.

## S9-3    Phage T4 Gene 32 Protein[19]

Phage T4 gene 32 protein, the first-discovered and best-studied of the single-strand binding proteins, has a molecular weight of 33,466 determined from its amino acid sequence.[20] Certain of its properties (pp. 282–287), prototypical for this class of binding proteins, deserve emphasis.

16. Kowalczykowski, S. C., Bear, D. G., and von Hippel, P. H. (1981) Enzymes 14, 373.
17. Hill, T. L., and Tsuchiya, T. (1981) PNAS 78, 4796.
18. Williams, K. R., and Konigsberg, W., personal communication.
19. Kowalczykowski, S. C., Bear, D. G., and von Hippel, P. H. (1981) Enzymes 14, 373.
20. Williams, K. R., LoPresti, M. B., Setoguchi, M., and Konigsberg, W. H. (1980) PNAS 77, 4614.

(i) _Domains_. As described for duplex binding proteins (Section S9-1), functionally discrete domains are responsible for binding to DNA, for cooperative interaction between molecules bound to DNA, for aggregation of unbound protein, and for interactions with functionally related proteins (e.g., DNA polymerase). At the amino-terminal end, residues 1–9 are clearly essential for cooperative binding to DNA. At the other end, residues 254–301 (which become more accessible to proteolysis when bound to DNA) may interact with other proteins in the replication complex and control the helix-destabilizing function of gene 32 protein.[21]

(ii) _Cooperative binding_. A single gene 32 protein molecule can bind to duplex DNA and to RNA, but it is the cooperativity of binding that confers the high specificity for single-stranded DNA and functional effectiveness.

## S9-4  Phage M13 Gene 5 Protein

The protein encoded by gene 5 of phage M13 and closely related filamentous phages (e.g., fd) is produced in huge amounts, about $10^5$ molecules per cell. By coating the viral strands, gene 5 protein prevents their being used as templates to produce more duplex replicative forms and prepares them for packaging into phage particles (p. 287).

The structure of the dimeric phage fd protein has been solved to 2.3-Å resolution, and cocrystals with oligonucleotides have been analyzed at lower resolution.[22] The unliganded protein has a 30-Å-long cleft that qualifies in size, shape, and amino acid composition as the DNA-binding site. Arrayed along the external edges of the groove are the aromatic residues that can stack on the nucleotide bases; binding of a tyrosine residue has also been observed by nuclear magnetic resonance.[23] Basic amino acids in the interior of the cleft may attract and bind the DNA backbone. The protein secondary structure consists solely of antiparallel $\beta$ sheets, one of which may participate in the nearest neighbor interactions for cooperative binding of the gene 5 protein to DNA. The nucleotide spacing when bound to the protein is 4.5 Å, nearly the same as that for the T4 gene 32 protein–DNA complex.

21. Williams, K. R., and Konigsberg, W. (1981) in _Gene Amplification and Analysis_ (Chirikjian, J., and Papas, T., eds.), Elsevier-North Holland, New York, vol. 2, p. 475.
22. McPherson, A., Jurnak, F. A., Wang, A. H. J., Molineux, I., and Rich, A. (1979) _JMB 134_, 379; McPherson, A., Jurnak, F., Wang, A., Kolpak, F., Rich, A., Molineux, I., and Fitzgerald, P. (1980) _Biophys. J. 32_, 155.
23. Garssen, G. J., Tesser, G. I., Schoenmakers, J. G. G., and Hilbers, C. W. (1980) _BBA 607_, 361.

## S9-5   *E. coli*, Phage T7, and Phage T5 Single-Strand Binding Proteins

### *E. coli* Single-Strand Binding Protein (SSB)[24]

*Structure.*   With the *ssb* gene cloned on a convenient plasmid,[25] the protein level can be amplified near 100-fold.[26] A simple ammonium sulfate precipitation of a lysate, as described for DNA polymerase III holoenzyme,[27] yielded per gram of cell paste 0.7 to 2.3 mg of SSB that is 97 percent pure on an SDS-polyacrylamide gel.[28] An abundant supply of SSB should now encourage its widespread use in biochemical studies and its intensive physicochemical investigation.

From the sequence of the *ssb* gene, its exact location and transcriptional orientation are known and the 177-amino-acid sequence of the protein (mol. wt. 18,873) has been deduced.[29] A protein domain with a high degree of secondary structure followed by one with very little secondary structure are anticipated and suggest functional homologies with the T4 binding protein (Section S9-3). Proteolytic cleavage studies show, as with the T4 protein, that removal of a large COOH-terminal domain does not prevent cooperative binding to DNA, but the complexes formed are kinetically more labile.[30]

*Binding properties.*[31]   Each of the four subunits of SSB binds an oligo dT about eight residues long, noncooperatively, as judged by quenching of intrinsic protein fluorescence. The SSB tetramer accommodates close to four $d(pT)_8$ molecules, two $d(pT)_{16}$, and only one $d(pT)_{30-40}$, the stoichiometry routinely observed under conditions of cooperative binding of SSB to DNA (p. 289). The binding constant increases 1000-fold when the oligonucleotide is enlarged from $d(pT)_{16}$ to $d(pT)_{30-40}$, a size sufficient to coil around the entire tetramer. The base composition of the DNA influences the stability of the complex, as shown by the far stronger interaction with oligo dT compared with oligo dA, suggesting that stretches of A residues in DNA might favor dissociation of SSB. Binding is remarkably insensitive to NaCl

---

24. Meyer, R. R., Glassberg, J., Scott, J. V., and Kornberg, A. (1980) *JBC* 255, 2897.
25. Sancar, A., and Rupp. W. D. (1979) *BBRC 90*, 123; Lebowitz, J., and McMacken, R., personal communication.
26. Chase, J. W., Whittier, R. F., Auerbach, J., Sancar, A., and Rupp, W. D. (1980) *N. A. Res. 8*, 3215.
27. McHenry, C., and Kornberg, A. (1977) *JBC 252*, 6478; Hübscher, U., Meyer, R., and Kornberg, A., unpublished results.
28. Soltis, D., personal communication; Lebowitz, J., and McMacken, R., personal communication.
29. Sancar, A., Williams, K. R., Chase, J. W., and Rupp, W. D. (1981) *PNAS 78*, 4274.
30. Williams, K. R., Spicer, E. K., LoPresti, M. B., Guggenheimer, R. A., and Chase, J. W., personal communication.
31. Krauss, G., Sindermann, H., Schomburg, U., and Maass, G. (1981) *B. 20*, 5346; Kowalczykowski, S. C., Lonberg, N., Newport, J. W., and von Hippel, P. M. (1981) *JMB 145*, 75; Schneider, R. J., and Wetmur, J. G. (1982) *B. 21*, 608.

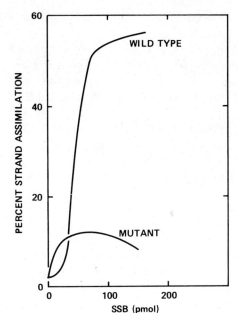

FIGURE S9-2
Stimulation of recA protein-catalyzed strand assimilation by single-strand binding protein (SSB) from wild-type compared to mutant (ssb⁻) cells. (Courtesy of Professor I. R. Lehman.)

in contrast to gene 32 protein whose affinity decreases by $10^7$ with a tenfold increase in NaCl concentration.

*Recombination and repair.* Participation of SSB in recombination, revealed in studies of the *ssb* mutant in vivo,[32] is seen dramatically in the enhancement of the strand-annealing activity of pure recA protein (Fig. S9-2).[33] By coating the single strands, SSB spares recA protein and promotes its efficient use; SSB very likely interacts with recA protein directly. The mutant SSB is inactive (Fig. S9-2). Those SSB interactions, as well as those seen between gene 32 protein and T4 DNA polymerase (p. 285), may be important in other SSB roles, such as interactions with proteins responsible for the priming and progress of DNA replication.

The sensitivity of *ssb* mutants to ultraviolet irradiation[34] is likely to be a deficiency in recombinational repair rather than excision-repair.[35] These strains are also defective in many of the functions induced in the SOS response (Section S16-2).

32. Glassberg, J., Meyer, R. R., and Kornberg, A. (1979) *J. Bact. 140,* 14.
33. McEntee, K., Weinstock, G. M., and Lehman, I. R. (1980) *PNAS 77,* 857.
34. Glassberg, J., Meyer, R. R., and Kornberg, A. (1979) *J. Bact. 140,* 14.
35. Whittier, R. F., and Chase, J. W. (1981) *MGG 183,* 341.

*Priming of replication.*   Several examples of SSB participation are known.

(i)   RNA polymerase priming of phage M13 single-strand replication depends on SSB for a unique start at the specific duplex region that marks the origin. SSB coating of the viral DNA destabilizes all but a few duplex regions, only one of which is recognized by the sigma subunit of RNA polymerase (Section S11-8).[36] The basis for specific recognition is puzzling in that the M13 origin region lacks promoter-like sequences. It also contains no particular sequence or secondary structure that distinguishes it from the other stem and loop regions in M13 (or in phage G4 and $\phi$X174 viral DNAs) that RNA polymerase also ignores.

(ii)   Protein n' depends on SSB for initiation of primosome assembly (Section S11-9) at a unique sequence of $\phi$X174 DNA[37] and at certain origins on each strand of ColE1 DNA.[38] Stoichiometric displacement of SSB by protein n' from coated DNA[39] may be one of the functions of protein n' as the engine that propels the primosome along DNA, displacing SSB transiently from its path.

(iii)   Protein n, whose essential participation in primosome assembly follows protein n' binding to DNA (Section S11-9), must form a complex with SSB in order to bind DNA.[40]

(iv)   Primase recognition of the origin for phage G4 complementary strand synthesis (p. 390) requires SSB coating of the viral strand.

In addition to directing initiation events and facilitating chain elongation (p. 290), SSB may also increase the fidelity of replication.[41]

Phage T7 and Phage T5 Binding Proteins

The gene for the T7 single-strand binding protein has been located[42] between genes 2 and 3 and named 2.5, but the relationship to a DNA replication mutation mapped close by is not known.[43] A gene 2.5 *amber* mutant, whose binding protein is 28 kdal instead of 32 kdal, is defective in replication, especially in a host with mutant SSB. A deficiency in DNA repair and recombination, more striking than that in replication, is observed with the T7 mutant even when the host is overproducing wild-type SSB. Such results indicate that the

36. Kaguni, J. M., and Kornberg, A. (1982) *JBC*, in press.
37. Shlomai, J., and Kornberg, A. (1980) *PNAS 77*, 799.
38. Nomura, N., Low, R. L., and Ray, D. S. (1982) *PNAS 78*, in press.
39. Shlomai, J., and Kornberg, A. (1980) *JBC 255*, 6789, 6794.
40. Low, R. L., Shlomai, J., and Kornberg, A. (1982) *JBC*, in press.
41. Kunkel, T. A., Meyer, R. R., and Loeb, L. A. (1979) *PNAS 76*, 6331.
42. Dunn, J. J., and Studier, F. W. (1981) *JMB 148*, 303; Araki, H., and Ogawa, H. (1981) *MGG 183*, 66.
43. North, R., and Molineux, I. J. (1980) *MGG 179*, 683.

host SSB is inadequate as a substitute for the phage protein especially in repair and recombination.

A binding protein encoded by phage T5 gene D5 is 29 kdal and produced in quantities near 2 percent of the cell protein. Cooperative binding to duplex DNA as well as noncooperative binding to single-stranded DNA[44] may relate to the requirement for this protein in viral DNA synthesis and transcription.

## S9-6 Eukaryotic Single-Strand Binding Proteins[45]

Eukaryotic single-strand binding proteins and their properties generally resemble the prokaryotic proteins but differ in two ways. They bind noncooperatively, relying perhaps on auxiliary functional proteins for greater effectiveness, and some are phosphorylated in a way that modulates their functions.

The 72-kdal protein encoded by adenovirus is the best understood of the eukaryotic single-strand binding proteins.[46] By binding to the terminus of duplex DNA[47] as well as to single strands, the adenoviral protein may function both in initiation of replication and in chain elongation (Section S15-4). A thermolabile mutant protein is temperature sensitive in viral DNA synthesis in vivo and in nuclear extracts. DNA synthesis can be restored to these extracts with a 44-kdal proteolytic fragment, containing the carboxy-terminal domain, as efficiently as with the intact wild-type protein.[48] The significance of the phosphorylation near the amino terminus is therefore still unclear.[49]

Herpes simplex virus encodes a DNA-binding and helix-destabilizing protein[50] that may resemble the adenoviral one.

A DNA-binding protein from the lily plant (*Lilium*) is found only in the germ cells and is presumed to be active in meiotic recombination.[51] A similar protein is also present in rat spermatocytes.[52] The lily protein complexes with single-stranded DNA even at 2 M

44. McCorquodale, D. J., Gossling, J., Benzinger, R., Chesney, R., Lawhorne, L., and Moyer, R. W. (1979) *J. Virol. 29*, 322; Rice, A. C., Ficht, T. A., Holladay, L. A., and Moyer, R. W. (1979) *JBC 254*, 8042.
45. Falaschi, A., Cobianchi, F., and Riva, S. (1980) *TIBS 5*, 154; Coleman, J. E., and Oakley, J. L. (1980) *Crit. Rev. Bioch. 7*, 247.
46. van der Vliet, P. C., Keegstra, W., and Jansz, H. S. (1978) *EJB 86*, 389; Schechter, N. M., Davies, W., and Anderson, C. W. (1980) *B. 19*, 2802; Kruijer, W., van Schaik, F. M. A., and Sussenbach, J. S. (1981) *N. A. Res. 9*, 4439.
47. Fowlkes, D. M., Lord, S. T., Linné, T., Pettersson, U., and Philipson, L. (1979) *JMB 132*, 163.
48. Ariga, H., Klein, H., Levine, A. J., and Horwitz, M. S. (1980) *Virol. 101*, 307.
49. Klein, H., Malzman, W., and Levine, A. J. (1979) *JBC 254*, 11051.
50. Powell, K. L., Littler, E., and Purifoy, D. J. M. (1981) *J. Virol. 39*, 894.
51. Hotta, Y., and Stern, H. (1979) *EJB 95*, 31.
52. Mather, J., and Hotta, Y. (1977) *Exp. Cell Res. 109*, 181.

NaCl and requires $Mg^{2+}$ or $Ca^{2+}$ for binding. Upon phosphorylation by a cAMP-dependent protein kinase, the lily protein will catalyze the melting of duplex DNA and the reannealing of single-stranded DNA.

## S9-7 Histone and Eukaryotic Packing Proteins

### Three-Dimensional Arrangement of Histones in the Nucleosome[53]

In histone-DNA crosslinking experiments,[54] reaction with dimethyl sulfate resulted in cleavage of the DNA at purine residues, followed by Schiff's base formation with neighboring proteins. When the procedure was applied to nucleosome core particles, $^{32}$P-labeled at the 5′ termini, a linear map of distances of the histones from the ends of the DNA was obtained.

The relationship of this map to the three-dimensional structure of the histone octamer was also determined by neutron scattering (Fig. S9-3) and electron microscopy.[55] Ordered aggregates of octamers were examined and images reconstructed to a resolution of 22 Å. The shape and symmetry of the octamer are apparent at this resolution, but not the outlines of the individual histones. The most prominent feature is a ridge that winds in a left-handed sense around the octamer and corresponds to the path followed by the DNA (Fig. S9-4). The locations of the individual histones along this path may be assigned on the basis of the map from chemical crosslinking. The three-dimensional data permit some clarification and simplification of the map, leading to the following linear arrangement of the histones along the DNA:

H2A–H2B–H4–H3–H3–H4–H2B–H2A

In three dimensions, this arrangement appears as a central, disk-shaped $(H3)_2(H4)_2$ tetramer, with an H2A–H2B dimer stacked on each face. The proposed central role of the tetramer in the nucleosome[56] is thus confirmed, and the dissociation of the octamer in solution by the loss of H2A–H2B dimers[57] is readily understood.

53. Kornberg, R. D., and Klug, A. (1981) *Sci. Amer.* 244 (no. 2), 52.
54. Mirzabekov, A. D., Shick, V. V., Belyavsky, A. V., and Bavykin, S. G. (1978) *PNAS* 75, 4184.
55. Klug, A., Rhodes, D., Smith, J., Finch, J. T., and Thomas, J. O. (1980) *Nat.* 287, 509.
56. Kornberg, R. D. (1974) *Science* 184, 868.
57. Thomas, J. O., and Kornberg, R. D. (1975) *PNAS* 72, 2626.

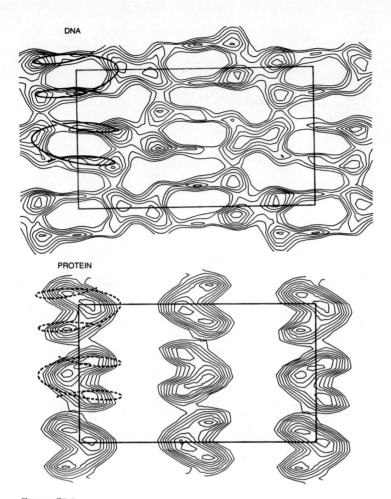

FIGURE S9-3
Neutron-scattering maps of a nucleosome core crystal. These maps
show separately the density of the DNA (top) and of the protein
(bottom). The DNA density correlates well with the projection of about
one and three-quarter turns of the DNA superhelix (solid line). The
DNA superhelix also appears to fit well (broken line) around the protein,
the histone octamer. (From Kornberg, R. D., and Klug, A. "The
Nucleosome." Copyright © 1981 by Scientific American, Inc. All rights
reserved.)

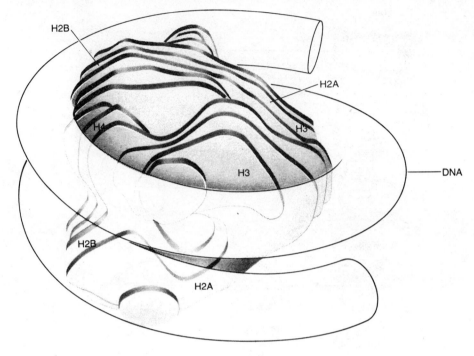

FIGURE S9-4
Model of the nucleosome core. The model was made by winding a tube simulating the DNA superhelix on a model of the histone octamer, which was built from a three-dimensional map derived from electron micrographs of the histone octamer. The ridges on the periphery of the octamer form a more or less continuous helical ramp on which a 146-nucleotide-pair length of DNA can be wound. The locations of individual histone molecules (whose boundaries are not defined at this resolution) are proposed here on the basis of chemical crosslinking data. (From Kornberg, R. D., and Klug, A. "The Nucleosome." Copyright © 1981 by Scientific American, Inc. All rights reserved.)

Nucleosome assembly might occur by the reverse of this reaction, H2A–H2B dimers being added to a complex of tetramer and DNA.

Nucleosome Assembly

Recent work has undermined what had been regarded as a basic tenet of nucleosome assembly—the linkage of histone and DNA synthesis. Rates of histone synthesis at various stages of the cell cycle can be measured by pulse-labeling synchronized cells with radioactive amino acids. In previous studies, histones were extracted from chromatin and found to become labeled only during S phase of the cell cycle; the conclusion was that histone and DNA synthesis are closely coupled.[58] In recent work, total cellular proteins were extracted and

58. Elgin, S. C. R., and Weintraub, H. (1975) *ARB* 44, 725.

the histones resolved by electrophoresis in two-dimensional gels. In contrast with the previous results, the histones were labeled to a similar extent throughout the cell cycle.[59] This suggests that histones are continually synthesized but assembled into chromatin only when new DNA appears during S phase. At other times during the cell cycle the newly synthesized histones form a pool, possibly associated with an acidic protein, the most abundant in *Xenopus* oocyte nuclei, that catalyzes chromatin assembly.[60] The factor for chromatin assembly in *Drosophila* embryos, thought initially to be topoisomerase I, is instead RNA;[61] in modulating the transfer of histones to DNA, the RNA can be replaced by other polyanions.[62]

High-Mobility-Group Proteins

Nonhistone chromosomal proteins, designated HMG (high mobility group) because of their electrophoretic mobility at low pH, fall into four groups in mammalian cells:[63] HMG 1 and HMG 2 are closely related and distinct from HMG 14 and HMG 17. Despite much effort to relate their abundance to differentiation and cellular proliferation, their physiological functions are still undefined.[64]

DNA Renaturation by Histones[65]

Either singly or in combination, histones from several sources promote the renaturation of single strands of denatured DNA. The reaction is complete in a minute and involves the histones in a stoichiometric rather than catalytic way. Since the reaction can be observed even with crude extracts of *Drosophila* embryos, it should provide a functional and sensitive assay for histones and other proteins that bind preferentially to duplex DNA.

## S9-8  Prokaryotic Packaging Proteins

Condensation and packaging of DNA in a variety of phages follow the same basic pattern, employing proheads, scaffolding proteins, and polyamines.[66] Formation of compact, toroidal DNA structures by

59. Groppi, V. E., Jr., and Coffino, P. (1980) *Cell 21*, 195; Seale, R. L. (1981) *B. 20*, 6432.
60. Laskey, R. A., Honda, B. M., Mills, A. D., and Finch, J. T. (1978) *Nat. 275*, 416; Mills, A. D., Laskey, R. A., Black, P., and De Robertis, E. M. (1980) *JMB 139*, 561.
61. Nelson, T., Wiegand, R., and Brutlag, D. (1981) *B. 20*, 2594.
62. Stein, A. (1979) *JMB 130*, 103; Stein, A., Whitlock, J. P., Jr., and Bina, M. (1979) *PNAS 76*, 5000.
63. Walker, J. M., Goodwin, G. H., and Johns, E. W. (1979) *FEBS Lett. 100*, 394; Walker, J. M., Gooderham, K., Hastings, J. R. B., Mayes, E., and Johns, E. W. (1980) *FEBS Lett. 122*, 264.
64. Seyedin, S. M., and Kistler, W. S. (1979) *JBC 254*, 11264; Seyedin, S. M., Pehrson, J. R., and Cole, R. D. (1981) *PNAS 78*, 5988.
65. Cox, M. M., and Lehman, I. R. (1981) *N. A. Res. 9*, 389.
66. Earnshaw, W. C., and Casjens, S. R. (1980) *Cell 21*, 319.

cobalt (III) hexaamine $[Co^{3+}(NH_3)_6]$, a simple polycation, may provide an instructive model reaction (Section S1-6). An abundant histone-like protein,[67] discovered in E. coli, may also help explain the beaded chromatin fibers in prokaryotes (p. 298) that were long thought to lack histones.

### H Protein[68]

A protein remarkably similar to histone H2A was recognized in fractionated extracts of E. coli because its binding to DNA inhibited replication, transcription, and other DNA-dependent reactions. The isolated protein, named H protein, is a dimer of 28-kdal polypeptides and binds 75 residues of single-stranded or duplex DNA. H protein is neutralized by rabbit antibodies against calf thymus H2A, and resembles that histone in amino acid composition, in DNA-binding features, and in its support of the reannealing of denatured DNA.[69]

### E. coli DNA-Binding Protein II

E. coli proteins identified with several activities related to DNA binding and named HU, HD, or NS prove to be the same and have been designated E. coli DNA-binding protein II;[70] protein I in this classification is SSB (Section S9-6). Protein II is made up of two related 90-amino-acid-long polypeptides (mol. wt. 9500) that differ in 30 percent of their residues.[71] The protein is a dimer or tetramer with the two polypeptides in a 1:1 ratio. Protein II binds SV40 DNA (p. 573) in the presence of a topoisomerase to form a nucleosome-like structure:[72] About 18 negative superhelical turns are introduced and the complex is condensed into a beaded fiber. An E. coli gene for a protamine-like protein, if expressed, may also contribute to DNA packaging.[73]

### The Bacterial Nucleoid

Binding proteins H and II may be abundant enough in E. coli to complex nearly half the genome. Possibly these and other fortuitously encountered proteins (p. 299),[74] augmented by still more that remain

67. Earnshaw, W. C., and Casjens, S. R. (1980) Cell 21, 319.
68. Hübscher, U., Lutz, H., and Kornberg, A. (1980) PNAS 77, 5097.
69. Cox, M. M. and Lehman, I. R. (1981) N. A. Res. 9, 389.
70. Geider, K., and Hoffmann-Berling, H. (1981) ARB 50, 233.
71. Mende, L., Timm, B., and Subramanian, A. R. (1978) FEBS Lett. 96, 395; Laine, B., Kmiecik, D., Sautiere, P., Biserte, G., and Cohen-Solal, M. (1980) EJB 103, 447; Rouvière-Yaniv, J., and Kjeldgaard, N. O. (1979) FEBS Lett. 106, 297.
72. Rouvière-Yaniv, J., Yaniv, M., and Germond, J.-E. (1979) Cell 17, 265.
73. Altman, S., Model, P., Dixon, G. H., and Wosnick, M. A. (1981) Cell 26, 299.
74. Geider, K., and Hoffmann-Berling, H. (1981) ARB 50, 233.

to be discovered, will constitute a set of bacterial histones. These together with polyamines, RNA, and nonhistone proteins may help account for the organization of the bacterial chromosome.

Despite the similarites of bacterial protein-DNA complexes to histones and nucleosomes, there are probably fundamental differences. The retention of DNA superhelicity after $\gamma$-irradiation of eukaryotic cells, measured by psoralen binding, is not observed in *E. coli*.[75] Instead, such treatment or a DNA gyrase inhibitor relaxes the supercoiling of *E. coli* DNA, implying that the torsional tension is not restrained, as in eukaryotic cells, by nucleosomal structures.

## S9-9   ATPases Dependent on DNA Binding

Enzymes with diverse functions, some considered in more detail under other headings, are collected in this section to emphasize a common theme: the use of the energy of hydrolyzing the $\beta-\gamma$ anhydride bond of a nucleoside triphosphate to favor a reaction on DNA. Helicases, for example, use ATP to move unidirectionally on DNA and destabilize duplex DNA in advance of a replication fork.

These enzymes belong to a large group of nucleoside triphosphatases (Table S9-1) whose reversible and energy-dependent conforma-

---

75. Sinden, R. R., Carlson, J. O., and Pettijohn, D. E. (1980) *Cell 21, 773.*

TABLE S9-1
Energy-transducing nucleoside triphosphatases[a]

| Class | Function | Example |
|---|---|---|
| Macromolecular biosynthesis | DNA replication and recombination | *ATPases:* gyrase, helicase, prepriming proteins, recA, recBC |
| | transcription | rho factor |
| | | mRNA splicing(?) |
| | translation | *GTPases:* IF-2, eIF-2, EF-Tu, EF-G, EF-1, EF-2, eRF |
| | | mRNA scanning |
| Biological movements and assembly | contraction | myosin, dynein ATPases |
| | transport | $Na^+$-$K^+$, $Ca^{2+}$ ATPases |
| | oxidative phosphorylation | $F_1$ ATPase |
| | cytoskeleton | microtubules microfilaments |
| Signal transfer | hormonal response | G-protein GTPase |
| | photoreception | light GTPase |

[a]Courtesy of Professor Y. Kaziro.

tional transitions are of prime importance for energy transduction in biological systems. These transitions depend on ligands noncovalently or covalently associated with the protein.

## Mechanism

As described for GTPase-driven protein synthetic reactions (p. 304), these ATPases, the helicases especially, operate on a "buy now, pay later" basis: Only after the protein-ATP complex has performed its function is ATP hydrolyzed to restore credit for a repetition of the cycle. A theoretical treatment[76] likens their behavior to the myosin-actin system in muscle. By contrast, the single-strand DNA-binding proteins do not consume ATP in moving on DNA and melting it. Their behavior resembles the steady state treadmilling of aggregates of tubulin in solution: As molecules are displaced from one end of a linear aggregate they reassociate with the other.

## Classification of DNA-Dependent ATPases

DNA-dependent ATPases, previously arranged alphabetically (p. 301), can now be grouped more meaningfully as follows: (i) helicases that destabilize the hydrogen-bonded base pairs to drive the unwinding of the double helix; (ii) proteins, auxiliary to a DNA polymerase, that use ATP energy to assist in polymerase binding and processivity; (iii) DNA-dependent ATPases with other identified functions: transcription (Section S7-8), topoisomerase (Sections S9-11, S9-12), recombination (recBC in Section S10-8), and restriction endonuclease (Section S10-6); and (iv) enzymes characterized only by an ATPase activity dependent on binding DNA but with no assigned functions (listed on p. 301); an additional 83-kdal entry will be mentioned below.

*Helicases.* See Table S9-2. Most helicases are processive in opening the duplex in advance of the replicating fork. Generally the enzyme binds and moves on the strand that serves as template for synthesis of the lagging strand (Figure S9-5). Movement therefore is opposite to that of lagging strand synthesis and has the $5' \rightarrow 3'$ direction of the template strand to which it binds. By contrast, rep protein (Section S9-10), while also processive, operates on the opposite strand; it binds to and moves on the $3' \rightarrow 5'$ oriented template strand in advance of synthesis of the leading strand. Thus rep protein acting on one parental strand, if assisted by a helicase acting on the other strand

---

76. Hill, T. L., and Tsuchiya, T. (1981) *PNAS 78*, 4796.

TABLE S9-2
Helicases[a]

| Helicase | Size, kdal | Stoichiometric | Processive | Direction | Copies/ Cell | Function |
|---|---|---|---|---|---|---|
| E. coli I[a] | 180 | yes | yes | 5′→3′ | 600 | unknown |
| E. coli II[a] | 75 | yes | no | 5′→3′ | 6000 | replication: E. coli, λ, ColE1 |
| E. coli III[b] | 20 | yes | yes | 5′→3′ | 20 | unknown |
| E. coli rep[c] | 65 | no[d] | yes | 3′→5′ | 50 | replication: φX174, M13 |
| E. coli primo-some[e] (n′, dnaB) | >500 | | yes | 5′→3′ | 20 | priming |
| E. coli recBC[f] | 268 | no[d] | | | | recombination |
| E. coli recA[g] | 38 | | | | 2000 | recombination |
| Phage T4 gene 41[e] | 58 | | | 5′→3′ | | priming |
| Phage T4 dda[h] | 56 | yes | no | | 6000 | unknown |
| Phage T7 gene 4[e] | 57, 66 | no[d] | yes | 5′→3′ | | priming |
| Lilium[i] | 130 | | | | | unknown |
| Human cell[j] | 110 | | | 5′→3′ | | unknown |

[a]Kuhn, B., Abdel-Monem, M., and Hoffmann-Berling, H. (1979) CSHS 43, 63; Abdel-Monem, M., and Hoffmann-Berling, H. (1980) TIBS 5, 128; Geider, K., and Hoffmann-Berling, H. (1981) ARB 50, 233.
[b]Yarranton, G. T., Das, R. H., and Gefter, M. L. (1979) JBC 254, 11997, 12001.
[c]Section S9-10
[d]Single-strand binding protein
[e]Section S11-10
[f]Section S10-8
[g]Section S16-4
[h]Purkey, R. M., and Ebisuzaki, K. (1977) EJB 75, 303.
[i]Hotta, Y., and Stern, H. (1978) B. 17, 1872.
[j]Cobianchi, F., Riva, S., Mastromei, G., Spadari, S., Pedrali-Noy, G., and Falaschi, A. (1979) CSHS 43, 639.

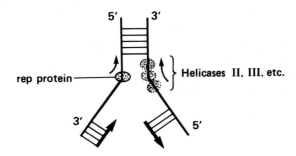

FIGURE S9-5
Polarity of helicase movements suggesting a possible joint action between rep protein moving on the 3′→5′ strand and another helicase (e.g., helicase II, III) moving on the 5′→3′ strand to melt the duplex in advance of a replicating fork.

(helicase II, III, etc.), might be even more effective in the unidirectional separation and unwinding of the parental duplex strands.

Rep protein as a helicase is necessary to drive the nicked, rolling-circle stage of $\phi$X174 and M13 replication (Section S9-10). However, the *rep* mutant, because of its leakiness or the availability of other helicases, is not significantly affected in replication of its own chromosome. Phage T4 gene 41-gene 61 proteins and phage T7 gene 4 protein function as both primases and helicases (Section S11-10); the *E. coli* primosome seems also to have helicase activity beyond its key role in priming (Section S11-10). Antibody-inhibition studies in vitro[77] suggest that helicase II functions in the replication of chromosomes, such as ColE1, phage $\lambda$, and *E. coli*, that initiate as closed circles.

*Auxiliary DNA polymerase proteins.*   Due to one or more of the many distinctive subunits of DNA polymerase III holoenzyme (Section S5-3), the energy of ATP hydrolysis is used to clamp the holoenzyme to the primer-template in a complex that thereafter maintains high processivity. The phage T4 auxiliary proteins (44–62–45) perform a similar function for the T4 polymerase.

A highly active DNA-dependent ATPase of 83 kdal has been separated from purified pol III holoenzyme preparations and distinguished from the $\tau$ subunit of nearly the same size.[78] This enzyme now joins the well-populated class of those that still lack a function.

## S9-10   Rep Protein: A Helicase

The stoichiometries and specificities of rep protein (rep) actions on single-stranded and duplex DNAs have improved our grasp of how rep functions as a helicase (Table S9-3). Rep can bind one molecule of ATP and a 20-residue stretch of single-stranded DNA.[79] In a ternary complex, the ATP is hydrolyzed and rep is promptly freed. A distinctive binding site for duplex DNA may be related to rep helicase action at a fork.

In the more physiological context of a $\phi$X174 rolling circle, where rep action is essential in vitro and in vivo, a stable isolatable complex contains the phage-encoded gene A protein and one molecule of rep.[80] Rep, oriented on a strand in the $3'{\rightarrow}5'$ direction,[81] moves processively to separate the duplex strands, consuming two molecules

77. Klinkert, M.-Q., Klein, A., and Abdel-Monem, M. (1980) *JBC 255*, 9746.
78. Meyer, R. R., Brown, C. L., and Rein, D. C., personal communication.
79. Arai, N., Arai, K., and Kornberg, A. (1981) *JBC 256*, 5287.
80. Arai, N., and Kornberg, A. (1981) *JBC 256*, 5294.
81. Yarranton, G. T., Das, R. H., and Gefter, M. L. (1979) *JBC 254*, 11997, 12002.

Table S9-3
Rep protein interactions with single-stranded DNA and replication fork compared[a]

| Property | SS DNA | Replication fork |
|---|---|---|
| ATPase activity | | |
| $K_m$ for ATP (dATP) ($\mu$M) | high (200) | low (25) and high (200) |
| Action on other rNTPs, dNTPs | yes | none or little |
| ATP analogs (relative $K_t$) | high | low |
| Mg$^{2+}$ optimum (mM) | broad (0.1–20) | sharp (5–10) |
| SSB | inhibits | activates |
| Gene A protein | no effect | required |
| Binding | | |
| Stoichiometry, one rep protein per | 20 residues | fork |
| Salt inhibition (100 mM NaCl) | yes | no |
| Effect of ATP or ADP | destabilizes | none |
| SSB | inhibits | no effect |
| Gene A protein | no effect | required |
| Kinetic mechanism | distributive | processive |

[a]From Arai, N., Arai, K., and Kornberg, A. (1981) *JBC* 256, 5287.

of ATP for each base pair melted. SSB is needed in this action, presumably to prevent the strands from reannealing. Possibly, rep and one of the helicases (Section S9-9) oriented on the strand with the 5′→3′ direction, coordinate their actions for more effective opening of the duplex.

# S9-11   Topoisomerases—Type I[82]

Type I topoisomerases, earlier known as the nicking-closing enzymes, make a break in one strand of a DNA duplex; type II topoisomerases make a break in both strands (Table S9-4).[83] Both types are ubiquitous. Typical type I enzymes are *E. coli* DNA topoisomerase I[84] (still earlier known as ω protein) and the eukaryotic nicking-closing enzymes.[85] In relaxation of DNA by the Type I

82. Cozzarelli, N. R. (1980) *Science* 207, 953; *Cell* 22, 327; Gellert, M. (1981) *ARB* 50, 879; Wang, J. C. (1981) *Enzymes* 14, 331.
83. Liu, L. F., Liu, C.-C., and Alberts, B. M. (1980) *Cell* 19, 697.
84. Brown, P. O., and Cozzarelli, N. R. (1979) *Science* 206, 1081.
85. Pulleyblank, D. E., Shure, M., Tang, D., Vinograd, J., and Vosberg, H.-P. (1975) *PNAS* 72, 4280; Rowe, T. C., Rusche, J. R., Brougham, M. J., and Holloman, W. K. (1981) *JBC* 256, 10354.

TABLE S9-4

Properties of type I and type II topoisomerases

| Property | Type I | Type II Gyrase | Eukaryotic[a] |
|---|---|---|---|
| DNA strands cleaved | one | two | two |
| Subunits | monomer | $\alpha_2\beta_2$ | dimer |
| ATP requirement | no | yes | yes |
| DNA-dep. ATPase | no | yes | yes |
| Gyration | no | yes | no |
| Relaxation | yes | yes[b] | yes |
| Catenation, knotting | yes[c] | yes | yes |

[a]*Drosophila*
[b]ATP not required
[c]Nick in one strand of duplex required

enzyme, one strand of the helix is passed through its complementary strand. The predicted "step-of-one" change in supercoiling value has been verified for *E. coli* DNA topoisomerase I.[86]

Insights into the mechanism of the type II topoisomerases (Section S9-12) helped clarify the type I mechanism and unified both in a rather simple and coherent pattern. In addition to the capacity of type I enzymes to make a transient single-strand break in a super-coiled duplex and thereby relax it, these topoisomerases also knot and catenate duplex DNA rings.[87] The requirement for a preexisting nick in at least one of the participating rings suggested that *E. coli* DNA topoisomerase I acts with high preference opposite the nick to open up the passage in the DNA that allows knotting and catenation. This has now been shown directly.[88] The facile catenation of rings of unrelated sequence suggests a new mechanism for all type I topoisomerase reactions, since previously it was thought that base pairing of passing and transiently broken segments was required (Fig. 9-12, p. 310). The new mechanism is directly analogous to the sign-inversion mechanism for type II enzymes except that it involves a transient enzyme-bridged single-strand break.

Although earlier isolation procedures of eukaryotic type I topo-isomerases produced a homogeneous 65- to 70-kdal enzyme in good yield, it is clear now that these were catalytically active proteolytic fragments of a 100- to 110-kdal polypeptide.[89]

86. Brown, P. O., and Cozzarelli, N. R. (1981) *PNAS 78*, 843.
87. Tse, Y.-C., and Wang, J. C. (1980) *Cell 22*, 269; Brown, P. O., and Cozzarelli, N. R. (1981) *PNAS 78*, 843.
88. Dean, F., and Cozzarelli, N. R., personal communication.
89. Liu, L. F., and Miller, K. G. (1981) *PNAS 78*, 3486; Wang, J. C., Gumport, R. I., Javaherian, K., Kirkegaard, K., Klevan, L., Kotewicz, M. L., and Tse, Y.-C. (1980) *ICN–UCLA Symposia 19*, 769.

Mutants of *E. coli* DNA topoisomerase I (*topA*) and their identity to *supX* of *S. typhimurium* provide clues to the function of this enzyme.[90] Deletion, nonsense, and insertion *topA* mutations cause only a modest decrease in growth rate of many strains, and thus topoisomerase I appears not to be essential. However, viability does seem to require extracistronic suppressors.[91] The consequences of *topA* mutations include diminished transposition, enhanced sensitivity to ultraviolet light, and suppression of weak, down-promoter mutants. The latter phenotype can be explained if the supercoil density of the chromosome is determined by a balance between the opposed actions of gyrase and topoisomerase I. Thus underwinding of the *topA* mutant DNA would facilitate RNA polymerase binding and overcome the promoter defect. Yet it seems unlikely that topoisomerase I acts only as a governor for gyrase. More plausibly, type I enzymes exploit their special ability to break and rejoin single strands and to act opposite single-strand breaks in other contexts.

Other Breakage-and-Reunion Enzymes

Although considered elsewhere, the actions of other breakage-and-reunion enzymes are noted here because they exhibit weak supercoil relaxing activity and their mechanisms likely resemble that of a type I topoisomerase. Unlike the classic topoisomerase, however, the break introduced by these enzymes is sealed only after their principal reaction has been completed. Gene A protein[92] of phage ΦX174 (Section S11-11) and the analogous gene 2 protein[93] of M13 (fd) are replication enzymes that break the viral strand of a supercoiled duplex at the unique origin of viral strand replication. Only after a full strand has been displaced does the reunion at the origin, regenerated by replication, take place to create a covalent single-stranded circle. Phage λ int protein,[94] a recombination enzyme, introduces a transient nick or break in one duplex and joins it to a nick in another duplex. The crucial feature of these several breakage-and-reunion enzymes is the transfer of single-stranded segments to produce a change in primary structure, a property also displayed by eukaryotic type I topoisomerases.[95] By contrast, the result of the classic type I topoisomerase action is generally a change in tertiary or quaternary structure of DNA.

90. Trucksis, M., and Depew, R. E. (1981) *PNAS 78*, 2164; Sternglanz, R., DiNardo, S., Voelkel, K. A., Nishimura, Y., Hirota, Y., Becherer, K., Zumstein, L., and Wang, J. C. (1981) *PNAS 78*, 2747; Trucksis, M., Golub, E. I., Zabel, D. J., and Depew, R. E. (1981) *J. Bact. 147*, 679.
91. Sternglanz, R., and Depew, R. E., personal communication.
92. Langeveld, S. A., van Arkel, G. A., and Weisbeek, P. J. (1980) *FEBS Lett. 114*, 269; van der Ende, A., Langeveld, S. A., Teertstra, R., van Arkel, G. A., and Weisbeek, P. J. (1981) *N. A. Res. 9*, 2037.
93. Meyer, T. F., and Geider, K. (1979) *JBC 254*, 12642.
94. Kikuchi, Y., and Nash, H. A. (1979) *PNAS 76*, 3760.
95. Been, M. D., and Champoux, J. J. (1981) *PNAS 78*, 2883.

## S9-12  Topoisomerases—Type II[96]

The mechanism of type II topoisomerases proved to be as simple as it was unexpected and will be illustrated below for DNA gyrase (Table S9-4). Type II enzymes include DNA gyrase,[97] the structurally related *E. coli* enzyme, topoisomerase II′,[98] T4 topoisomerase (Section S14-6),[99] and the ATP-dependent relaxing enzymes from a variety of eukaryotic sources such as *Drosophila*,[100] *Xenopus laevis*,[101] and HeLa cells.[102]

Type II topoisomerases change the linking number of DNA not by acting on DNA twist but by changing the writhing. Gyrase binds to a DNA molecule in such a way that two DNA segments cross, forming two supercoiled loops, one positive and the other negative (Fig. S9-6). By passing one of the DNA segments through a reversible break in the other, the sign of the positive supercoil is inverted and thus the mechanism is referred to as "sign inversion." Two negative supercoils are thus introduced in each catalytic step. Reversal of the scheme removes supercoils two at a time. The demonstration of the "step-of-two" change in supercoiling provided the first critical test for the model.[103]

Catenation and knotting of intact duplex DNA rings was anticipated by the sign-inversion mechanism and vividly demonstrates the ability of topoisomerases to pass DNA segments through each other.[104] Catenanes result when the crossing DNA segments come from different DNA rings, and knotting can be viewed as intramolecular catenation. A function of topoisomerases may be the resolution of catenanes and knots into constituent rings, inasmuch as catenanes are in vitro products of recombination[105] and replication,[106] and even phage virion DNA is found to be knotted.[107]

Catenation, unlike all other topoisomerase reactions, requires a polyvalent cation such as spermidine or histone. By neutralizing

96.  Gellert, M. (1981) *Enzymes 14*, 345.
97.  Liu, L. F., Liu, C.-C., and Alberts, B. M. (1980) *Cell 19*, 697; Brown, P. O., and Cozzarelli, N. R. (1979) *Science 206*, 1081; Mizuuchi, K., Fisher, L. M., O'Dea, M. H., and Gellert, M. (1980) *PNAS 77*, 1847.
98.  Brown, P. O., Peebles, C. L., and Cozzarelli, N. R. (1979) *PNAS 76*, 6110; Gellert, M., Fisher, L. M., and O'Dea, M. H. (1979) *PNAS 76*, 6289.
99.  Liu, L. F., Liu, C.-C., and Alberts, B. M. (1980) *Cell 19*, 697.
100. Hsieh, T.-S., and Brutlag, D. (1980) *Cell 21*, 115.
101. Baldi, M. I., Benedetti, P., Mattoccia, E., and Tocchini-Valentini, G. P. (1980) *Cell 20*, 461.
102. Miller, K. G., Liu, L. F., and Englund, P. T. (1981) *JBC 256*, 9334.
103. Brown, P. O., and Cozzarelli, N. R. (1979) *Science 206*, 1081; Liu, L. F., Liu, C.-C., and Alberts, B. M. (1980) *Cell 19*, 697; Mizuuchi, K., Fisher, L. M., O'Dea, M. H., and Gellert, M. (1980) *PNAS 77*, 1847.
104. Cozzarelli, N. R. (1980) *Cell 22*, 327.
105. Pollock, T. J., and Nash, H. A. (1980) *ICN–UCLA Symposia 19*, 953; Kolodner, R. (1980) *PNAS 77*, 4847; Reed, R. R. (1981) *Cell 25*, 713.
106. Sundin, O., and Varshavsky, A. (1980) *Cell 21*, 103.
107. Liu, L. F., Perkocha, L., Calendar, R., and Wang, J. C. (1981) *PNAS 78*, 5498.

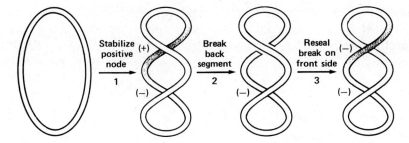

FIGURE S9-6
"Sign-inversion" mechanism for DNA gyrase. (Courtesy of Professor N. Cozzarelli.)

nearly all negative charge on DNA, polycations compact DNA into aggregates in which the high local DNA concentration drives the intermolecular reaction of catenation.[108] Dispersal of the aggregates by removal of the polycation or addition of competing mono- or divalent cations promotes decatenation.

## DNA Gyrase

A new purification procedure for E. *coli* gyrase employs affinity chromatography on novobiocin-Sepharose.[109] It has an $\alpha_2\beta_2$ tetrameric structure in which the $\alpha$ protomer is encoded by the gene *gyrA* (formerly called *nalA*) and the $\beta$ protomer by *gyrB* (formerly *cou*). The *Micrococcus luteus*[110] and E. *coli*[111] enzymes share this structure; functional hybrid enzymes can be formed.[112]

A closely related E. *coli* topoisomerase activity, called topoisomerase II', has been reconstituted from the $\alpha$ protomer and an equally abundant 50-kdal polypeptide that appears to be a fragment of the larger $\beta$ protomer.[112] Topoisomerase II' differs from gyrase in lacking negative supercoiling and ATPase activities. Furthermore, its capacity to relax positive as well as negative supercoils qualify it as an alternative for gyrase as a swivelase in replication.

The reversible cleavage of DNA required for topoisomerization is accompanied by covalent attachment of the revealed 5'-phosphoryl

108. Kreuzer, K. N., and Cozzarelli, N. R. (1980) *Cell 20*, 245; Krasnow, M. A., and Cozzarelli, N. R. (1982) *JBC 257*, 2687.
109. Staudenbauer, W. L., and Orr, E. (1981) *N. A. Res. 9*, 3589.
110. Klevan, L., and Wang, J. C. (1980) *B. 19*, 5229.
111. Sugino, A., Higgins, N. P., and Cozzarelli, N. R. (1980) *N. A. Res. 8*, 3865.
112. Brown, P. O., Peebles, C. L., and Cozzarelli, N. R. (1979) *PNAS 76*, 6110; Gellert, M., Fisher, L. M., and O'Dea, M. H. (1979) *PNAS 76*, 6289.

groups to a tyrosine residue in each $\alpha$ protomer.[113] Cleavage is not dictated by a unique DNA sequence, but some sequences occur preferentially at the cleavage site. A consensus sequence for many sites is:

N denotes any base; the arrows in the sequence denote cuts staggered by four base pairs on opposite strands.[114] The cleavage site is contained within about 140 base pairs of DNA wrapped on the surface of the enzyme as revealed by their protection from nuclease digestion.[115] A detailed model for supercoiling unites these results with the sign-inversion mechanism (Fig. S9-7).

Topoisomerases in Replication

Topoisomerases could use their capacities to perform several functions in the replication of closed circular duplex DNA. Supercoiling by gyrases could facilitate binding of enzymes and factors at the origin of initiation of a cycle of chromosome replication. By relaxing positive supercoiling that builds up in advance of the replication fork, a topoisomerase could act as a swivel. A gyrase could in addition facilitate unwinding of the fork by introducing negative supercoiling. Finally, when a round of replication is completed, catenated products could be separated by topoisomerase actions. Recent evidence to support topoisomerase functions in replication comes from in vivo studies of bacteria,[116] plasmids,[117] phage T7,[118] and phage T4.[119]

## S9-13    Covalent Protein-DNA Complexes

Topoisomerases perform their strand breakage-reunions with energy conservation through a covalent bond between the protein and a

113. Morrison, A., and Cozzarelli, N. R. (1979) *Cell 17*, 175; Tse, Y.-C., Kirkegaard, K., and Wang, J. C. (1980) *JBC 255*, 5560.
114. Morrison, A., Brown, P. O., Kreuzer, K. N., Otter, R., Gerrard, S. P., and Cozzarelli, N. R. (1980) *ICN–UCLA Symposia 19*, 785.
115. Morrison, A., and Cozzarelli, N. R. (1981) *PNAS 78*, 1416; Kirkegaard, K., and Wang, J. C. (1981) *Cell 23*, 721; Fisher, L. M., Mizuuchi, K., O'Dea, M. H., Ohmori, H., and Gellert, M. (1981) *PNAS 78*, 4165.
116. Orr, E., Fairweather, N. F., Holland, I. B., and Pritchard, R. H. (1979) *MGG 177*, 103; Filutowicz, M. (1980) *MGG 177*, 301; Snyder, M., and Drlica, K. (1979) *JMB 131*, 287; Filutowicz, M., and Jonczyk, P. (1981) *MGG 183*, 134.
117. Itoh, T., and Tomizawa, J. (1978) *CSHS 43*, 409; Taylor, D. E., and Levine, J. G. (1979) *MGG 174*, 127.
118. De Wyngaert, M. A., and Hinkle, D. C. (1979) *J. Virol. 29*, 529.
119. Liu, L. F., Liu, C.-C., and Alberts, B. M. (1979) *Nat. 281*, 456; McCarthy, D. (1979) *JMB 127*, 265.

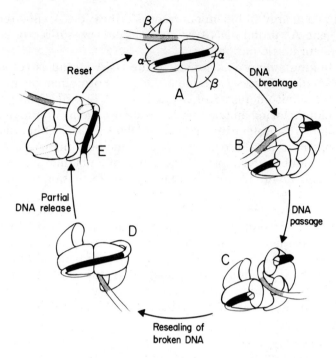

FIGURE S9-7
Model for supercoiling of DNA by gyrase. (A) A gyrase molecule with a section
of a circular duplex DNA wrapped in a right-handed superhelix. The solid DNA
represents the approximately 50-bp central region completely protected from DNase I
and contains the site of double-strand breakage at its middle. The stippled DNA is
to be passed through the transient gap. The enzyme contains a hole (concealed in (A))
that is exposed by opening of the molecule about a hinge between the $\alpha$ protomers
(as in (B)). To effect supercoiling, a right-handed DNA loop in (A) is converted by
sign inversion into a left-handed one in (D) as follows: (B) DNA breakage and opening
of the enzyme; (C) passing of DNA through the gap by exchange of the stippled
DNA from one $\beta$ protomer to the other; and (D) closing of the enzyme and resealing
of the broken DNA. (E) The DNA is reset to complete the cycle; the resealed DNA
section is released, followed by reopening of gyrase to allow escape of the passed
DNA section. The cycle could be keyed to the binding and hydrolysis of ATP as
follows. With gyrase in the closed conformation (A or D), interaction of the stippled
DNA with a $\beta$ promoter could allow $\beta$ to bind an ATP molecule. This in turn could
trigger a conformational change, i.e., opening of the gyrase molecule (as in (B) or (E)).
ATP hydrolysis could then stimulate release of DNA from a $\beta$ protomer and reclosing
of gyrase. (Courtesy of Professor N. Cozzarelli.)

phosphoryl residue at the break. The linkage is to the 5′-phosphoryl
in all but the eukaryotic type I topoisomerases[120] in which it is to the
3′. The protein-DNA complex is known to be the active intermediate
in single-strand transfer by the rat liver topoisomerase;[121] phosphodiester bond formation occurs as phosphotyrosine is broken

120. Champoux, J. J. (1977) *PNAS 74*, 3800; Prell, B., and Vosberg, H.-P. (1980) *EJB 108*, 389.
121. Been, M. D., and Champoux, J. J. (1981) *PNAS 78*, 2883.

down in an energy-independent transesterification. Analogously the $\phi$X174 gene A* protein-DNA complex circularizes directly without any cofactor participation.[122]

The linking amino acid joined to the DNA strand is tyrosine in all the topoisomerases examined: the *gyrA* protomer of *M. luteus* gyrase,[123] topoisomerase I of *M. luteus* and *E. coli*,[123] and the rat liver nicking-closing enzyme.[124] Tyrosine is also the amino acid that links polio virus protein VPg to the polio RNA (p. 317). Other instances of esterification of a tyrosine side chain include linkage to a single uridylyl residue in the allosteric regulation of glutamine synthetase,[125] and to phosphate in the neoplastic transformation of cells by viruses.[126]

Other covalent protein-DNA linkages are found in the virions and replicative intermediates of phage $\phi$29[127] and adenovirus.[128] Here the linkage is between a serine side chain and the nucleotide (dAMP in phage $\phi$29 or dCMP in adenovirus) that represents the 5'-phosphoryl termini of both strands of the linear DNA duplex. These covalent intermediates could provide a novel priming mechanism for initiation of replication by these genomes (Sections S14-8, S15-4).

---

122. Eisenberg, S., and Finer, M. (1980) *N. A. Res. 8*, 5305.
123. Tse, Y.-C., Kirkegaard, K., and Wang, J. C. (1980) *JBC 255*, 5560.
124. Champoux, J. J. (1981) *JBC 256*, 4805.
125. Shapiro, B. M., and Stadtman, E. R. (1968) *JBC 243*, 3769.
126. Bishop, J. M. (1981) *Cell 23*, 5.
127. Salas, M., and Viñuela, E. (1980) *TIBS 5*, 191.
128. Kelly, T. J., Jr., Challberg, M. D., and Desiderio, S. V. (1980) *ICN–UCLA Symposia 19*, 91.

# S10

# Deoxyribonucleases

## Abstract

Were there a Citation Index for enzymes, the deoxyribonucleases[1] would likely now be in first place. Their multiplicity and remarkable diversity have charged the explosive growth of genetic engineering. For example, a newcomer, *Bal*31 nuclease, a single-strand endonuclease that acts on the slightly frayed ends of a duplex, has proven to to be an ideal reagent for a finely graded reduction of DNA length, while still leaving the DNA suitable for blunt-end ligation to other DNA duplexes. Restriction endonucleases continue to be discovered, and a third major class of these enzymes has now been established.

---

1. Linn, S. (1981) *Enzymes 14*, 121.

An inactive *Neurospora* polypeptide is the source of four or more distinctive endonucleases derived by proteolytic processing. Nucleases remain key elements in repair and recombination processes. Certain incision endonucleases recognize AP (apurinic or apyrimidinic) sites and, acting also as glycosylases, first detach the altered base before incising the phosphodiester bond at one or the other side of it.

Ribonucleases, except for brief descriptions of RNaseH, are not covered in *DNA Replication* or in this Supplement. Volume 15 of *The Enzymes*, which will appear at about the time of this Supplement, focuses on the enzymes that synthesize, degrade, and modify RNA.

## S10-2–S10-5   Exonucleases[2] and Endonucleases

### Bal31 Nuclease

For its effectiveness in progressively shortening both strands of a duplex,[3] *Bal*31 nuclease, from the salt-water bacterium *Altermonas espejiana,* has become a favored reagent in genetic engineering. As a single-strand specific nuclease, its properties are similar to those of enzymes obtained from fungi, including *Aspergillus* (S1 nuclease), *Neurospora crassa,* and *Ustilago maydis.* The uniqueness of *Bal*31 nuclease resides in its additional capacity to hydrolyze the frayed ends of a linear duplex DNA dispersively. Both strands of the duplex are chewed away without introducing significant strand breaks away from the termini. Such DNA fragments are suitable for blunt-end ligation to other DNA duplexes.

### Exonuclease III

Nucleotides with sulfur substituted in the $\alpha$-phosphate ($\alpha$-phosphorothioates) can be polymerized by DNA polymerase I and sealed by DNA ligase but are not substrates for removal from chain ends by exonuclease III.[4] Thus exonuclease action on a duplex DNA with a phosphorothioate at only one 3'-strand end releases that strand intact if degradation of the other strand is taken to completion. Alternatively, if coupled with S1 nuclease, both strands of the duplex can be degraded specifically from one end. A comparable inability of the $3' \rightarrow 5'$ exonuclease activity of DNA polymerase I to degrade chains

2. Weiss, B. (1981) *Enzymes 14,* 203.
3. Legerski, R. J., Hodnett, J. L., and Gray, H. B., Jr. (1978) *N. A. Res. 5,* 1445.
4. Putney, S. D., Benkovic, S. J., and Schimmel, P. R. (1981) *PNAS 78,* 7350.

terminated in a phosphorothioate blocks the proofreading capacity
of the polymerase (Sections S4-10, S4-11).

### *Neurospora* Nucleases[5]

Intensive study of what may have seemed to be an undistinguished
single-strand endonuclease from *N. crassa* has disclosed novel features that may have wide metabolic significance.[6] An inactive, precursor polypeptide (95 kdal) is processed proteolytically to produce
several other forms. These include a single-strand specific endonuclease, a single-strand specific exonuclease, a mitochondrial
DNase, and a periplasmic DNase. All forms of the enzyme are anti-
genically related. A mutant strain appears to lack a protease responsible for processing an integral mitochondrial inner membrane form
of the enzyme to one loosely associated with the mitochondrial membrane, and a cytosol form to one secreted by lysosome-like vesicles.
The mutant is characterized by abnormal meiosis, abnormal mitotic
recombination, and a high spontaneous mutation rate, but not by
ultraviolet- or x-ray-induced mutagenesis. When the details of the
various enzyme forms, functions, and interconversions become
known, the importance of proteases in regulating the activities and
physiological functions of nucleases will be better appreciated.

### Mammalian DNases

The number and variety of these nucleases increase as a direct function of the investigative attention given them. Introduction of a
standardized nomenclature[7] may be premature however, because
adequate information is not yet available either to associate multiple activities with discrete enzymes or to recognize identical enzymes derived from diverse sources.

## S10-6, S10-7    Restriction Endonucleases[8]

In addition to the accepted groupings of types I and II nucleases, a
new class with intermediate properties warrants designation as
type III (Table S10-1).

---

5. Lehman, I. R. (1981) *Enzymes 14*, 193.
6. Fraser, M. J., Chow, T. Y.-K, and Kafer, E. (1980) in *DNA Repair and Mutagenesis in Eukaryotes* (Generoso, W. M., Shelby, M. D., and de Serres, F., eds.) Plenum Press, New York, p. 63.
7. Hollis, G. F., and Grossman, L. (1981) *JBC 256*, 8074.
8. Yuan, R. (1981) *ARB 50*, 285; Endlich, B., and Linn, S. (1981) *Enzymes 14*, 137; Wells, R. D., Klein, R. D., and Singleton, C. K. (1981) *Enzymes 14*, 157.

TABLE S10-1
Restriction endonucleases

| Characteristics | Type I | Type II | Type III |
|---|---|---|---|
| Examples | EcoB | EcoRI | EcoP1 |
| Restriction and methylation[a] | single enzyme | separate enzymes | single enzyme |
| Subunits | distinctive (3) | simple dimer | distinctive (2) |
| Restriction factors | | | |
|   Essential | AdoMet,[b] ATP, $Mg^{2+}$ | $Mg^{2+}$ | ATP, $Mg^{2+}$ |
|   Stimulatory | | | AdoMet |
| Host-specificity site | hyphen of 6–8 nucleotides | symmetry (twofold) | no symmetry |
| Cleavage site | random; >1 kb from host-specificity site | at host-specificity site | 24–26 bp to unique side of host-specificity site |
| Enzymatic turnover | no | yes | yes |
| DNA translocation | yes | no | no |
| Methylation factors | | | |
|   Essential | AdoMet | AdoMet | AdoMet |
|   Stimulatory | ATP, $Mg^{2+}$ | | ATP, $Mg^{2+}$ |
| ATPase | extensive | none | none |
| Methylation at host-specificity site | yes | yes | yes |
| Overall complexity | great | modest | great |

[a]A type I enzyme performs one or the other; a type III performs both simultaneously; a type II endonuclease is distinct from the methylase.
[b]AdoMet, S-adenosylmethionine

## Type I Endonucleases

The recognition sequence for EcoK has been determined[9] and can be compared to that of EcoB:

$$
\begin{array}{l}
\text{EcoB} \quad 5'\text{ - T G }\overset{*}{\text{A}}\text{ N}_1\text{ N}_2\text{ N}_3\text{ N}_4\text{ N}_5\text{ N}_6\text{ N}_7\text{ N}_8\text{ T G C T} \\
\qquad\qquad\text{A C T N N N N N N N N A C G A - }5' \\
\text{EcoK} \quad 5'\text{ - A }\overset{*}{\text{A}}\text{ C N}_1\text{ N}_2\text{ N}_3\text{ N}_4\text{ N}_5\text{ N}_6\text{ G T G C} \\
\qquad\qquad\text{T T G N N N N N N C A C G - }5'
\end{array}
$$

N represents nonspecific nucleotides and the asterisk, sites of methylation.

The ability of the EcoK and EcoB enzyme complexes to exchange active subunits in vivo suggests an evolutionary relationship. Going

9. Kan, N. C., Lautenberger, J. A., Edgell, M. H., and Hutchinson, C. A., III (1979) JMB 130, 191; Brooks, J. E., and Roberts, R. J. (1982) N. A. Res. 10, 913.

from the EcoB recognition sequence to that of EcoK, the outermost specific nucleotides are lost and two new specific base pairs are adopted in the inner, nonspecific "hyphen" region. The locations of the methylated adenine residues remain constant, however. This relationship should be recalled when recognition (host-specificity) sequences are compared in other contexts.

## Type II Endonucleases

Many more of these enzymes continue to be discovered and novel specificities are disclosed among them. An annual updating of their tabulation seems essential.[10] Detailed studies[11] of the EcoRI restriction and modification enzymes showed the two enzymes to be rather dissimilar in their contacts with the DNA substrate, even though they recognize the same sequence. A dimeric endonuclease interacts with the 6-bp sequence to form a complex with twofold symmetry, appropriate for cleavage of both strands. The methylase binds as a monomer asymmetrically transferring methyl groups one at a time and dissociating from the DNA between transfers.

## Type III Endonucleases

This class, resembling type I in complexity, includes the EcoP1, EcoP15, and HinfIII endonucleases.[12] These enzyme systems are characterized by two genes and two subunits (of approximately 100 and 75 kdal) that restrict and modify. The smaller subunit can carry out the modification reaction without the larger subunit.

The recognition (host-specificity) sequence of these enzymes[13] are:

| 5'-A G A C C | 5'-C A G C A G | 5'-C G A A T |
|---|---|---|
| T C T G G-5' | G T C G T C-5' | G C T T A-5' |
| EcoP1 | EcoP15 | HinfIII |

Because methylation is confined to adenine, the EcoP1 and EcoP15 sequences can be modified on only one strand. The HinfIII recognition sequence is also methylated on only one of its strands. It is

10. Roberts, R. J. (1980) N. A. Res. 8, r63; (1982) in The Nucleases (Linn, S., and Roberts, R. J. eds.) CSHL, in press; (1982) N. A. Res. 10, r117.
11. Modrich, P. (1979) Quart. Rev. Biophys. 12, 315; Lu, A.-L., Jack W. E., and Modrich, P. (1981) JBC 256, 13200.
12. Bickle, T. A. (1982) in The Nucleases (Linn, S., and Roberts, R. J., eds.) CSHL, in press.
13. Bächi, B., Reiser, J., and Pirrotta, V. (1979) JMB 128, 143; Hadi, S. M., Bächi, B., Shepherd, J. C. W., Yuan, R., Ineichen, K., and Bickle, T. A. (1979) JMB 134, 655; Piekarowicz, A., Bickle, T. A., Shepherd, J. C. W., and Ineichen, K. (1981) JMB 146, 167.

still unclear how the daughter duplex containing the unmethylated parental strand can be protected from restriction directly following replication.

Each of the type III endonucleases cleaves DNA at a point 24–26 base pairs to the right of the host-specificity sequence (shown above), and each leaves a 2- or 3-nucleotide single-stranded stretch at the 5′ termini. These restriction enzymes require ATP, but unlike the type I enzymes, do not cleave it. S-adenosylmethionine (AdoMet) stimulates nuclease activity but is not required. In the absence of ATP each endonuclease can act specifically as the appropriate modification methylase.

## S10-8 Nucleases in Repair and Recombination

### uvrABC Endonuclease of *E. coli*

The structural genes of this major incision system for DNA repair have each been cloned; the *uvrA*, *B*, and *C* gene products have been purified to homogeneity and have masses of 114, 84, and 70 (or 65) kdal, respectively.[14] The uvrA and C proteins bind to single-stranded DNA, as does the uvrB protein in the presence of the uvrA protein;[15] the uvrA protein has ATPase activity. No subunit appears to have detectable endonuclease activity in the absence of the other two. All three genes seem to be controlled by the *recA lexA* regulatory circuit, but there is no evidence that the induced levels of these proteins are needed for an optimal rate of incision in UV-irradiated cells. It also remains unclear whether the three gene products act as a complex or individually and whether they act in a concerted or in a sequential manner.

### Phage T4 UV Endonuclease and *M. luteus* UV Endonuclease

Both phage T4 UV endonuclease (T4 endoV) and *M. luteus* UV endonuclease proceed by a glycosylase-AP endonuclease mechanism[16] (Fig. S10-1). Each also acts on DNA containing a simple AP site, as

---

14. Sancar, A., Wharton, R. P., Seltzer, S., Kacinski, B. M., Clarke, N. D., and Rupp, W. D. (1981) *JMB 148*, 45; Sancar, A., Clarke, N. D., Griswold, J., Kennedy, W. J., and Rupp, W. D. (1981) *JMB 148*, 63; Sancar, A., Kacinski, B. M., Mott, D. L., and Rupp, W. D. (1981) *PNAS 78*, 5450; Sharma, S., Ohta, A., Dowhan, W., and Moses, R. E. (1981) *PNAS 78*, 6033.
15. Kacinski, B. M., and Rupp, W. D. (1981) *Nat. 294*, 480.
16. Radany, E. H., and Friedberg, E. C. (1980) *Nat. 286*, 182; Haseltine, W. A., Gordon, L. K., Lindan, C. P., Grafstrom, R. H., Shaper, N. L., and Grossman, L. (1980) *Nat. 285*, 634; Gordon, L. K., and Haseltine, W. A. (1980) *JBC 255*, 12047; Seawell, P. C., Smith, C. A., and Ganesan, A. K. (1980) *J. Virol. 35*, 790; Demple, B., and Linn, S. (1980) *Nat. 287*, 203.

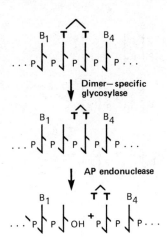

FIGURE S10-1
Cleavage mechanism of
phage T4 (T4 endoV) and *M.
luteus* UV endonucleases.
Following the cyclobutane-
dimer specific N-glycosylase
cleavage, incision by an AP
(apurinic-apyrimidinic)
endonuclease takes place.

shown in the second stage of the reaction in Figure S10-1. Both
activities reside in the small T4 endoV protein, the product of the T4
*denV* gene.[17]

The unusual pyrimidine-dimer nucleotide formed in vitro by these
enzymes on UV-irradiated DNA has also been observed in *M. luteus*
cells and in phage T4-infected cells,[18] but not in detectable amounts
in uninfected *E. coli* or in cultured human fibroblasts;[19] the latter
contain little, if any, pyrimidine-dimer glycosylase activity. The *denV*
gene, cloned into *E. coli recAuvrA* or *recAuvrB* mutants, com-
plements a deficiency of these mutants in incision at pyrimidine
dimers.[20]

## AP Endonucleases

Acting generally only on duplex DNA, AP endonucleases fall into
two classes, depending upon the location of incision (Fig. S10-2).[21]
The combined actions of an enzyme from each class can remove
deoxyribose 5-phosphate from DNA containing AP sites. The single
nucleotide gap generated by this double digestion can be filled with
comparatively little "repair synthesis." DNA polymerase I is also
effective in removing 3'-deoxyribose termini and displacing strands
with 5'-deoxyribose 5'-P termini.

17. Nakabeppu, Y., and Sekiguchi, M. (1981) *PNAS 78*, 2742; Warner, H. B., Christensen, L. M.,
    and Persson, M.-L. (1981) *J. Virol. 40*, 204; McMillan, S., Edenberg, H. J., Radany, E. H.,
    Friedberg, R. C., and Friedberg, E. C. (1981) *J. Virol. 40*, 211.
18. Radany, E. H., and Friedberg, E. C. (1982) *J. Virol. 41*, 88.
19. Demple, B., and Linn, S. (1980) *Nat. 287*, 203; La Belle, M., and Linn, S. (1982) *Nat.*, submitted.
20. Lloyd, R. S., and Hanawalt, P. C. (1981) *PNAS 78*, 2796.
21. Mosbaugh, D. W., and Linn, S. J. (1980) *JBC 255*, 11743; (1982) *JBC 257*, 575.

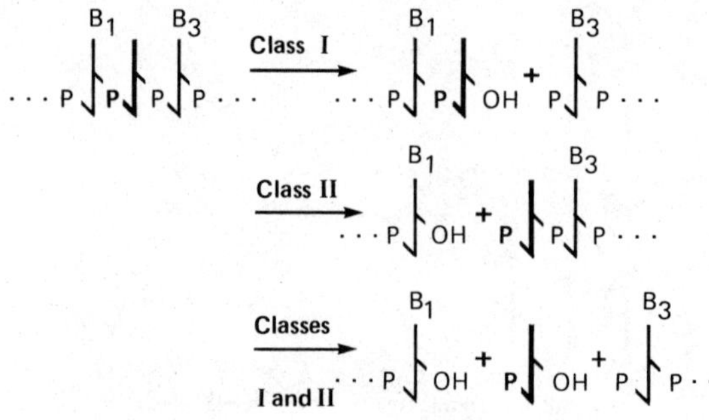

FIGURE S10-2
Cleavage mechanisms of class I and class II AP
endonucleases. (Courtesy of Professor S. Linn.)

The endonuclease activity associated with *E. coli* exonuclease III
(p. 324), also known as endonuclease VI, specifically recognizes only
AP sites on DNA. Earlier observations of an endonuclease II that recognizes alkylated nucleotides were likely due to the combined actions of endonuclease VI and DNA glycosylase.

The majority of AP endonucleases are in class II; the most noteworthy are *E. coli* endonuclease IV and endonuclease VI (exonuclease III), HeLa AP endonuclease, and human placental AP endonuclease. All bacterial class I enzymes are associated with DNA
glycosylase activity: the *M. luteus* and phage T4 UV endonucleases
and *E. coli* endonuclease III, the latter a thymine-glycol (dihydrothymine) glycosylase. A class I AP endonuclease from cultured
human fibroblasts, as mentioned above, has no associated DNA glycosylase.

### The *E. coli* recBC Enzyme[22]

Also known as exonuclease V, the *E. coli* recBC enzyme is regulated
in its action upon single-stranded DNA by recA protein[23] and by
SSB.[24] The regulation of the enzyme's degradation of linear duplex
DNA is uncertain, especially in vivo where susceptible duplex termini are normally absent or inaccessible. Novel looped structures
are observed microscopically among recBC reaction intermediates[25]
and may help in defining the enzyme's role in recombination.

22. Muskavitch, K. M. T., and Linn, S. (1981) *Enzymes 14*, 233.
23. Williams, J. G. K., Shibata, T., and Radding, C. M. (1981) *JBC 256*, 7573.
24. Williams, J. G. K., Shibata, T., and Radding, C. M. (1981) *JBC 256*, 7573; Rosamond, J.,
Telander, K. M., and Linn, S. J. (1979) *JBC 254*, 8646.
25. Muskavitch, K. M. T., and Linn, S. (1980) *ICN–UCLA Symposia 19*, 901; (1982) *JBC 257*, 2641;
Taylor, A., and Smith, G. R. (1980) *Cell 22*, 447.

# S11

# Replication Mechanisms and Operations

## Abstract

Semidiscontinuous replication of the double helix appears to be a basic design of nature: continuity of synthesis of the leading strand and discontinuous synthesis of the lagging strand. Because DNA polymerases, equipped as holoenzymes with numerous accessory proteins, are highly processive (rarely dissociate from the DNA template), discontinuities in synthesis of the leading strand, if found, may be viewed as exceptions to this general rule.

Another rule of wide generality is that RNA primes the starts of discontinuous DNA synthesis. In two instances, however, a viral protein appears to start a DNA chain by fixing the initiating nucleotide. The complexity of the array of proteins (seven proteins, twenty polypeptides) that prime chain starts in replication of the *E. coli* chromosome is reduced by their organization into a locomotive-like primosome that also moves processively to advance the replication fork. An adequate description of the structures and functions of the many prepriming proteins and primases now available in homogeneous form from several sources would require a separate chapter in a new edition of *DNA Replication*. The possibility that DNA polymerase is disposed as a twin assembly that carries out concurrent replication of both strands evokes an image of the polymerases, primosome and additional helicases integrated into a single unit, the replisome (larger than a ribosome) that creates and advances the fork of semiconservative replication. The high fidelity of the process is assured largely by proofreading operations built into polymerase action and also by repair units that monitor the product.

Two neglected aspects of replication are gaining attention: the start and the termination of a chromosome cycle. Discovery of a potent enzyme system that recognizes the unique, 245-base-pair origin of the *E. coli* chromosome and initiates bidirectional replication from that point, exposes the process to biochemical analysis for the first time. When the identities of the proteins that recognize and manipulate the origin sequence become known, the regulatory mechanisms that operate the switch for replication should come into view. Topological problems that confront the termination of replication of a duplex circle and separation of the twin products can be solved by one of the variety of topoisomerases, but it may be naive to expect that the segregation of chromosomes, so clearly linked to cell division, will escape further complexities.

## S11-3 Semidiscontinuous Replication

The mechanism of semidiscontinuous replication based on studies in vivo and in vitro (Chapter 11) has gained in credibility from enzyme studies in several systems, but has also been challenged by others.[1] Activation of *E. coli* DNA polymerase III holoenzyme for each start is slow and consumes seconds, but once started is rapid (near 1000 nucleotides per second) and proceeds with unlimited processivity (Section S5-3). Thus, continuous synthesis of the leading strand seems to be the most efficient and least complicated mechanism. Discovery of the organized primosome (Section S11-10) as a highly processive priming device operating unidirectionally on the strand with 5′→3′ polarity (Fig. S11-11, below) suits it well for discontinuous synthesis of the lagging strand. The need for repeated activations of the polymerase for these starts may be circumvented

---

1. Ogawa, T., and Okazaki, T. (1980) *ARB 49*, 421.

(Section S11-15). However, in SV40 DNA, nascent small fragments are produced at the replicating fork on both strands in isolated nuclei[2] and in vivo.[3] Ribonucleotides at the 5′ ends of these fragments are compelling evidence that discontinuous synthesis can occur.

Validity of the semidiscontinuous replication mechanism should be judged by its generality rather than universality. Many factors may operate to interrupt continuous synthesis and favor fresh starts. Low temperatures imposed experimentally to trap nascent fragments may favor discontinuous synthesis as may other circumstances, such as an imbalance of replication proteins or dNTPs, unreplicatable lesions in the template or secondary template structures that resist melting. In such cases, failure of continuous synthesis to keep pace with fork opening might create opportunities for primed starts of discontinuous synthesis.

## S11-4   Uracil Incorporation into DNA

The relative pool sizes of dUTP and dTTP govern the incorporation of uracil into DNA, and the effectiveness of the uracil N-glycosylase repair mechanism determines its persistence in DNA. Ordinarily, dUTPase activity leaves no significant dUTP pool but deficiency of the enzyme, as in *dut* mutants, or reduction of the dTTP pool, as by methotrexate treatment, may elevate dUTP to relatively high levels (Section S2-8).

DNA fragments produced by excision of uracil from DNA in post-replication repair in E. *coli* can be distinguished from nascent, RNA-primed replication fragments by retention of a 5′-phosphate terminus after alkaline hydrolysis.[4] Resistance of these fragments to degradation by spleen exonuclease has led to an additional demonstration that uracil-derived fragments are not significant in number except in dUTPase mutants.

## S11-5   RNA Priming of Chains in Discontinuous Replication

The starts of DNA chains in the discontinuous events of replication (Section S11-3), as well as the initiation of complementary strands on single-stranded genomes (Sections 11-8, 11-9, 11-10), are made by covalent extensions of short RNA transcripts. A survey of the primer RNAs (Table S11-1) reveals variations in their size and composition.

2. Närkhammar-Meuth, M., Eliasson, R., and Magnusson, G. (1981) *J. Virol. 39*, 11.
3. Närkhammar-Meuth, M., Kowalski, J., and Denhardt, D. T. (1981) *J. Virol. 39*, 21.
4. Machida, Y., Okazaki, T., Miyake, T., Ohtsuka, E., and Ikehara, M. (1981) *N. A. Res. 9*, 4755.

TABLE S11-1
Primer RNAs for discontinuous DNA replication[a]

| Genome | Chain length of RNA linked to DNA fragments | Primary products |
|---|---|---|
| T7 phage | 1 to 5 | pppApCpC/A(pN)$_{1-2}$[b] (N: A and C rich) |
| T4 phage | 1 to 5 | pppApC(pN)$_3$ |
| E. coli | 1 to 3 | not determined |
| $\phi$X174 phage | 1 to 5[c] | [d] |
| Sea urchin | 1 to ~8 | (p)ppA/G(pN)$_7$ |
| Polyoma, SV40 (isolated nuclei) | ~10 | pppA/G(pN)$_{\sim 9}$ |
| Animal cells | ~9 | |

[a]Adapted from tabular summary furnished by Professor T. Okazaki.
[b]The complementary template sequence that signals the primer is:
3'-pCpTpGpG/Tp—; G/T signifies that either G or T is found in this position.
[c]Of primers linked to the complementary strand in the SS→RF conversion in vitro, 39% were 1 residue long, 18% 2 residues, and 83% between 1 and 5 residues; comparable values for the reaction in vivo were: 1 residue 48%, 2 residues 23%, and 1 to 5 residues 92%.
[d]The 5'-terminal residue is A in about 80% of the primers in vitro.

The length of a primer can vary from one to several residues, the size distribution centering near five nucleotides with the T4 and T7 systems and rather sharply near ten nucleotides with primers in animal cells. The initiating nucleotide is generally ATP. Sites of initiation and termination are influenced by the coupling of helicase, primase, and polymerase actions as well as the sequence and secondary structure of the template.

## S11-8    Initiation by RNA Polymerase

The nature of RNA polymerase function in the initiation of replication of the E. coli chromosome (Section S11-15), plasmids (Section S14-9), and single-stranded phages (below) is now better understood.

### Specificity of Priming Phage M13 Replication[5]

RNA polymerase primes the replication of M13 DNA but not $\phi$X174 DNA in vivo and in crude extracts of E. coli. However, purified preparations of RNA polymerase did not distinguish between the two templates covered with single-strand binding protein (p. 385). In the

---

5. Kaguni J. M., and Kornberg, A. (1982) JBC 257, 5437.

course of preparing homogeneous RNA polymerase, loss of specificity and decreased priming activity can result from procedures that remove the σ subunit. Specificity is restored and priming activity increased upon the addition of σ subunit to core RNA polymerase. Priming of replication depends on a very limited amount of transcription, presumably confined to the unique sequence in M13 DNA that represents the origin of complementary strand synthesis. No transcription is observed on $\phi$X174 DNA comparably covered by binding protein (SSB). Thus, the specific priming of M13 DNA replication in *E. coli* depends on recognition of an origin sequence by an RNA polymerase holoenzyme with a functional σ subunit. Priming specificity for M13 DNA replication therefore provides a sensitive and simple test for the activity of the σ subunit of *E. coli* RNA polymerase, even though the M13 origin region lacks the sequences characteristic of RNA polymerase promoters.

## S11-9   Prepriming for Initiation by Primase: The Preprimosome as a Mobile Replication Promoter in *E. coli*[6]

The complex of proteins that initiates conversion of the single-stranded $\phi$X viral circle to the duplex form is important and remarkable for several reasons:

(i)   The complex includes the components that operate at the replicating fork of the *E. coli* chromosome to initiate the nascent fragments in the discontinuous synthesis of the lagging strand (Fig. S11-11, below).

(ii)   The *preprimosome* complex contains one molecule each of protein n′, protein n, and dnaB protein, and likely one or more of the other proteins (protein i, protein n″, and dnaC protein) that are essential for its assembly and function. All six proteins have been purified to near homogeneity. The locomotive-like preprimosome, when augmented by primase, becomes a *primosome* (Section S11-10). The primosome moves processively with the replicating fork to prime the repeated initiations of the nascent strands.

(iii)   Assembly and function of the primosome are key events in both viral and host chromosomal replication (Fig. S11-1). However, the composition of the primosome acting at the origin of the *E. coli* chromosome may differ or be more exacting than the one that is effective with the viral genome.

---

6. Arai, K., Arai, N., Shlomai, J., Kobori, J., Polder, L., Low, R., Hübscher, U., Bertsch, L., and Kornberg, A. (1981) *Prog. N. A. Res. 26*, 9; Arai, K., Low, R., Kobori, J., Shlomai, J., and Kornberg, A. (1981) *JBC 256*, 5273.

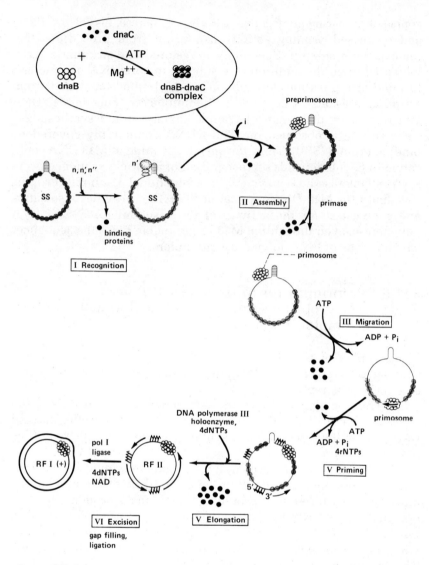

FIGURE S11-1
Scheme for assembly and migration of the primosome and the stepwise
displacement of SSB in the $\phi$X SS→RF reaction.

Protein n'[7]

This 76-kdal polypeptide recognizes a 55-nucleotide fragment from
$\phi$X174 DNA even in the presence of single-strand binding protein
(Fig. S11-2).[8] This fragment is located within the untranslated region

7. Shlomai, J., and Kornberg, A. (1980) *JBC 255*, 6789; 6794.
8. Shlomai, J., and Kornberg, A. (1980) *PNAS 77*, 799.

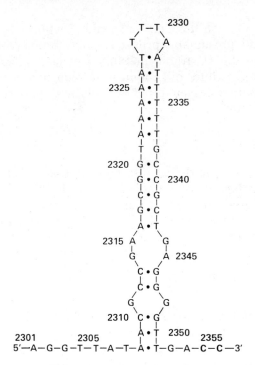

FIGURE S11-2
Potential secondary structure in the
recognition locus of protein n'. This proposed
stem-loop structure is based on a calculation
of the free-energy contributions of
base-paired regions and loops within the
55-nucleotide fragment of $\phi$X174 DNA that
fully supports ATP hydrolysis by protein n'.

between genes *F* and *G*, a map location analogous to that of the unique origin for synthesis of the complementary strand of phage G4 DNA. Within the fragment is a sequence that can form a stable, 44-nucleotide hairpin structure. This duplex may provide the signal for protein n' to initiate the prepriming events that lead to the start of $\phi$X174 complementary strand replication. Specificity for this duplex region is shown by lack of n' binding to other duplex DNAs and SSB-covered single-stranded DNAs.

Binding of one protein n' molecule is manifested by ATPase (or dATPase) activity and initiates assembly of the primosome. The capacity of protein n' to displace SSB and use ATP energy to propel the primosome rapidly along DNA to other locations on the $\phi$X genome accounts for the lack of a unique initiation site of the $\phi$X complementary strand in vivo and in vitro.

Protein n' also displays ATPase and primosome assembly functions at two sites on the ColE1 plasmid (Section S14-9).[9] One site, on the L strand, presumably primes lagging strand synthesis, and the other site, on the H strand, possibly primes replicative conjugal strand transfer. Failure of protein n' to recognize a site at or near the E. coli origin (e.g., denatured oriC plasmid) may reflect the need for additional factors.

## Protein n[10]

This 12-kdal polypeptide is a dimer in solution. It is heat- and acid-resistant, and N-ethylmaleimide-sensitive. Its binding to DNA depends on a direct interaction with SSB; as many as 30 monomers can be bound to a $\phi$X circle. However, in the assembly of the primosome guided by protein n', only one monomer is incorporated and serves in the subsequent binding of dnaB protein.

## Protein n''[11]

This 17-kdal polypeptide shares the heat- and acid-resistant properties of protein n, but is also resistant to N-ethylmaleimide. The low abundance of protein n'' and its presence as a contaminant in other replication protein preparations made its assay and isolation difficult. Protein n'' is essential for priming replication, but its function, stoichiometry, and fate are still unknown.

## Protein i[12]

Protein i is a trimer of 22-kdal polypeptides. Incorporation into the preprimosome occurs at a stage involving the dnaB-dnaC complex and after the actions of proteins n', n, and n''.

## dnaB Protein

The abundance of this hexamer of 50-kdal polypeptides is amplified over 200-fold in cells with runaway plasmids bearing the *dnaB* gene.[13] Large quantities of crystalline protein are easily obtained (320 mg from 1.5 kg of cell paste),[14] inviting studies of its structure and function. As a DNA-dependent ATPase (or rNTPase),[15] it had

9. Zipursky, S. L., and Marians, K. J. (1980) *PNAS 77*, 6521; Nomura, N., and Ray, D. S. (1980) *PNAS 77*, 6566; Nomura, N., Low, R. L., and Ray, D. S., *PNAS 79*, in press; Böldicke, T. W., Hillenbrand, G., Lanka, E., and Staudenbauer, W. L. (1981) *N. A. Res. 9*, 5215.
10. Low, R. L., Shlomai, J., and Kornberg, A. (1982) *JBC 257*, in press.
11. Liao, T., and Kornberg, A., unpublished observation.
12. Arai, K., McMacken, R., Yasuda, S., and Kornberg, A. (1981) *JBC 256*, 5281.
13. Yasuda, S., Kobori, J., Arai, K., and Kornberg, A., unpublished observation.
14. Arai, K., Yasuda, S., and Kornberg, A. (1981) *JBC 256*, 5247.
15. Reha-Krantz, L. J., and Hurwitz, J. (1978) *JBC 253*, 4043; 4051; Arai, K., and Kornberg, A. (1981) *JBC 256*, 5253; 5260.

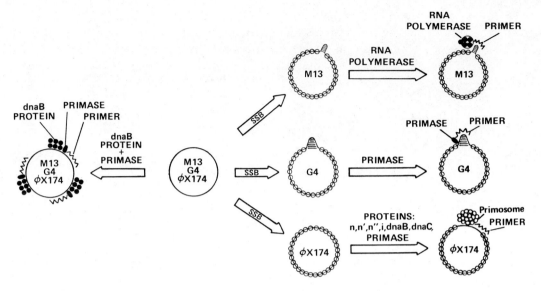

FIGURE S11-3
Nonspecific (general) priming by dnaB protein and primase on uncoated DNA and the specific priming systems for M13, G4, and φX174 DNAs coated with SSB.

been regarded as responsible for the mobility of the prepriming complex (p. 386). Instead, this ATP-dependent function is likely performed by protein n'; dnaB protein uses ATP to alter DNA secondary structure for primase action. To pursue the locomotive analogy for the primosome, protein n' has the role of the engine (conveniently equipped with a cowcatcher to displace SSB on the track) and dnaB protein acts as the engineer.

The basic function of dnaB protein is illustrated in general (non-specific) priming on uncoated DNA (Fig. S11-3).[16] ATP, or a non-hydrolyzable ATP analog, induces a high affinity complex that forms or stabilizes a DNA secondary structure exploitable by primase. Upon ATP hydrolysis, the low affinity ADP-dnaB protein complex dissociates (Fig. S11-4).[17] For specific priming of replication on SSB-coated DNA (Fig. S11-3), dnaB protein must be part of the primosome complex. The complex, again formed even with a nonhydrolyzable ATP analog, is stable, mobile, and processive.

The dnaB protein has a multifaceted structure (Fig. S11-5).[18] In addition to specific sites for ATPase activity, primase, dnaC protein (see below), single-stranded DNA, and duplex DNA, allosteric effects of ATP and dATP on these complexes are also prominent.[19] Binding

16. Arai, K., and Kornberg, A. (1981) *JBC 256*, 5267.
17. Arai, K., and Kornberg, A. (1981) *JBC 256*, 5260.
18. Arai, K., and Kornberg, A. (1981) *JBC 256*, 5253.
19. Arai, K., and Kornberg, A. (1981) *JBC 256*, 5260.

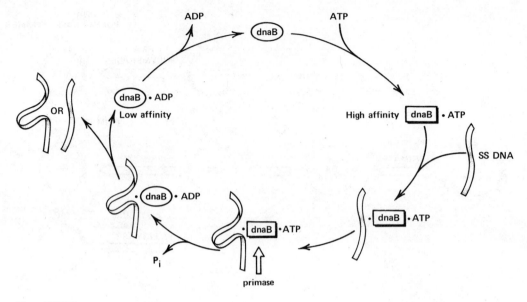

FIGURE S11-4
Distributive reaction mechanism for ATP-induced conformational changes of dnaB protein acting on DNA as a "replication promoter." In effect, dnaB protein engineers secondary structural features that favor primase binding and transcriptional synthesis of primer.

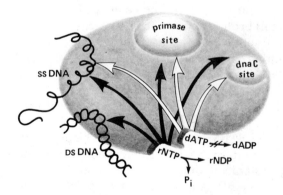

FIGURE S11-5
Hypothetical structural and functional relationships of the monomeric unit of dnaB protein. Omitted are the binding sites to the other dnaB units needed to form the native hexameric structure. The arrows emanating from the rNTP and dATP sites represent allosteric effects on the sites to which they are directed.

of ATP (and ADP) can also be manifested by stabilization of the protein to heating[20] and more dramatically by activation of mutant forms that otherwise appear inert.[21]

Limited proteolytic cleavage of dnaB protein reveals domains in its organization.[22] With removal of fourteen amino acids from the N-terminal region, replicative interactions are lost, but ATPase activity and binding of ATP and single-stranded DNA are preserved in a 30-kdal C-terminal polypeptide.

Hybrid oligomers of wild-type and mutant dnaB protein display a variety of properties in vitro and in vivo. A recent example[23] demonstrated the lability of the hybrid oligomer in certain of the groP mutants that fail to support the growth of phage λ (p. 549). The dnaB protein defect can be alleviated either by an alteration in the λ P gene (whose product interacts functionally with dnaB protein) or by prophage P1 subunits that form a more stable heteromultimer with the host protein.

### dnaC Protein and the B-C Complex

Cloning of the *dnaC* gene and a 50-fold overproduction of the 30-kdal dnaC protein has made its isolation and characterization possible.[24] A complex of six molecules of dnaC protein formed with a dnaB protein hexamer makes the dnaC protein resistant to N-ethylmaleimide. It is the B-C complex, with participation of protein i, that transfers one molecule each of dnaB protein and protein i to the developing primosome. The fate of the dnaC protein in this assembly remains to be determined.

## S11-10  Initiation by Primases and Primosomes[25]

RNA priming of DNA chain synthesis is nearly universal and generally the transcriptional responsibility of a specialized RNA polymerase, called primase. With further studies of known primases and discovery of new examples, their inclusion in complex assemblies of

20. Arai, K., Yasuda, S., and Kornberg, A. (1981) *JBC 256*, 5247.
21. Günther, E., Mikolajczyk, M., and Schuster, H. (1981) *JBC 256*, 11970.
22. Arai, K., Arai, N., and Nakayama, N. (1982) in *The Future of Nucleic Acid Research*, AMBO Symposium (Japan), in press.
23. Günther, E., Lanka, E., Mikolajczyk, M., and Schuster, H. (1981) *JBC 256*, 10712.
24. Kobori, J., and Kornberg, A., unpublished results.
25. Arai, K., and Kornberg, A. (1981) *PNAS 78*, 69; Arai, K., Low, R. L., and Kornberg, A. (1981) *PNAS 78*, 707; Low, R. L., Arai, K., and Kornberg, A. (1981) *PNAS 78*, 1436; Arai, K., Low, R., Kobori, J., Shlomai, J., and Kornberg, A. (1981) *JBC 256*, 5273; Benz, E. W., Jr., Remberg, D., and Hurwitz, J. (1982) *Enzymes 15*, 155.

priming and replicative proteins, called primosomes or replisomes, may become the rule.

### E. coli Primase (dnaG Protein) and the Primosome

The *dnaG* gene forms an operon with *rpoD*, the gene for the σ subunit of RNA polymerase. Cloning of the operon by overcoming a powerful terminator can amplify the level of primase about 50- to 100-fold and make its isolation in useful quantities more feasible.[26] Actions of primase can be studied in conjunction with dnaB protein on un-coated DNA (Fig. S11-3), and on SSB-coated DNA directly with phage G4 DNA or as part of a primosome with phage φX174 or ColE1 supercoiled DNAs.

*Primase with G4 DNA.*   Primase recognizes and transcribes a hair-pin region in the viral DNA of phage G4 and a similar region in several closely related phages (p. 390).[27] How many molecules of primase bind and transcribe what particular sequence is still uncer-tain. Although a one to one primase:G4 DNA complex may exist,[28] a 2:1 complex likely represents the functional form.[29] Such a com-plex may ultimately explain how transcription coupled to replica-tion is limited to a few residues, but when uncoupled can be ex-tended to 29 residues.

*Primosome.*[30]   On SSB-coated φX174 DNA, on supercoiled ColE1 plasmid, and presumably at replicating forks of the *E. coli* chromo-some, primase action is physically and functionally linked to dnaB protein in a primosome. By its association with the protein compo-nents of a preprimosome (Section S11-9), primase is bound to the viral DNA or the lagging strand template and moves processively in its 5′→3′ orientation, a direction opposite to primer and DNA chain elongation (Fig. S11-6). Unlike the gene 4 protein of phage T7 (below), primase relies on a primosomal component, protein n′, for propul-sion and possibly helicase action.

The primosome once assembled on an invading φX single-stranded circle or appropriated from the host replicating fork remains bound even after the circle becomes a covalently closed, supercoiled du-plex (RFI) (Fig. S11-7). Conservation of the primosome facilitates the next stage of RF replication by (i) directing initiation of the rolling

26. Wold, M. S., and McMacken, R., personal communication.
27. Sims, J., Capon, D., and Dressler, D. (1979) *JBC* 254, 12615.
28. Benz, E. W., Jr., Reinberg, D., Vicuna, R., and Hurwitz, J. (1980) *JBC* 255, 1096; Sims, J., and Benz, E. W., Jr. (1980) *PNAS* 77, 900.
29. Stayton, M., and Kornberg, A., unpublished observation.
30. Arai, K., and Kornberg, A. (1981) *PNAS* 78, 69; Arai, K., Low, R. L., and Kornberg, A. (1981) *PNAS* 78, 707; Low, R. L., Arai, K., and Kornberg, A. (1981) *PNAS* 78, 1436; Arai, K., Low, R., Kobori, J., Shlomai, J., and Kornberg, A. (1981) *JBC* 256, 5273.

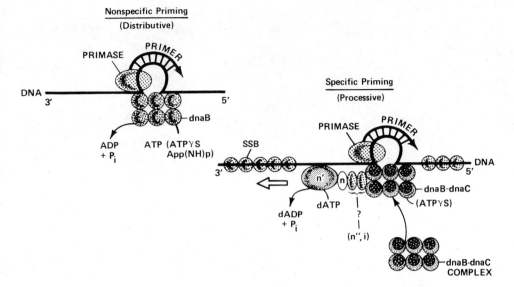

FIGURE S11-6
Proposed mechanisms for nonspecific (general) and specific priming (see Fig. S11-3). ATP analogs (e.g., ATPγS) not hydrolyzed by dnaB protein support complex formation with uncoated DNA (upper left) and primosome assembly on coated φX174 DNA (lower right). For primosome movement, ATP (or dATP) hydrolysis is essential and presumably the function of protein n'. The stoichiometry of the primosome subunits is known only for the dnaB-dnaC complex, protein n', and protein n.

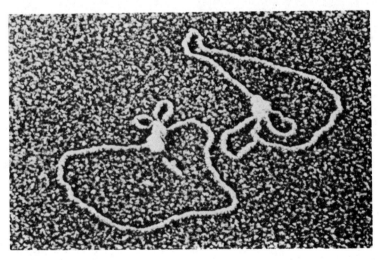

FIGURE S11-7
Electron micrograph of the primosome bound to covalently closed φX174 duplex replicative form. These forms invariably contain a single primosome with one or two associated small DNA loops. (Courtesy of Professor J. Griffith.)

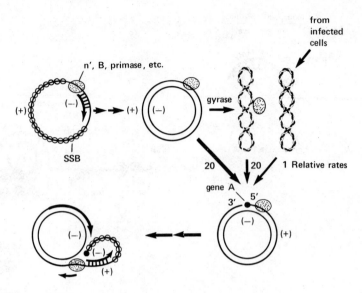

FIGURE S11-8
Conservation of the primosome throughout the first two stages of the
$\phi$X174 DNA replicative cycle. B represents dnaB protein.

circle to the unique site of gene A protein cleavage, (ii) performing at
the replicating fork in priming complementary strand synthesis, and
(iii) serving, perhaps, as a helicase (Fig. S11-8). The parental RF
bearing the primosome may be the sole template for replication,
while the numerous supercoiled progeny RF are produced for transcription.

The integrity and functional capacity of the primosome-RFII complex depend on experimental isolation conditions. For example,
the complex isolated by sucrose-gradient sedimentation appears to
have retained proteins that are lost by gel filtration. Retention of
functional primase in the primosome is especially variable. These
findings raise the possibility that additional proteins present in the
primosome in vivo have been lost during isolation and have gone
undetected because the need for them was not observed in the assays
of the $\phi$X174 SS→RF conversion. Assays provided by the more exacting oriC plasmid replication system (Section S11-15) may lead to
their discovery.

Phage, Plasmid, and Eukaryotic Primases

*Phage T7 gene 4 protein.*[31]    The protein may consist of two species
of polypeptide chains of 57 and 66 kdal. Both species appear to have

31. Tabor, S., Engler, M. J., Fuller, C. W., Lechner, R. L., Matson, S. W., Romano, L. J., Saito, H.,
Tamanoi, F., and Richardson, C. C. (1981) *ICN–UCLA Symposia 22*, 387.

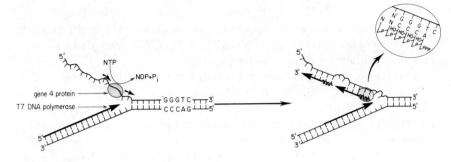

FIGURE S11-9
Helicase and primase actions of phage T7 gene 4 protein. (Courtesy of Professor C. C. Richardson.)

the enzymatic activities of gene 4 protein. Based on nucleotide sequence analysis, gene 4 has two initiation sites for protein synthesis in the same reading frame at positions that would give rise to the two polypeptides observed.

Gene 4 protein is both a helicase and a primase (Fig. S11-9). It binds specifically and stably to single-stranded DNA in the presence of a nucleoside triphosphate, and is then translocated on the DNA in a 5' to 3' direction[32] in a reaction coupled to hydrolysis of nucleoside triphosphates. The $K_m$ for the preferred substrate, dTTP, is $4 \times 10^{-4}$ M. Nonhydrolyzable nucleoside triphosphates such as $\beta,\gamma$-methylene-dTTP induce tighter binding but inhibit both primase and helicase activities. Nucleoside triphosphate hydrolysis thus appears to be essential for unidirectional movement of the gene 4 protein to unwind DNA and to bring itself to a recognition site for its primase activity.

Helicase activity of gene 4 protein is observed during synthesis on duplex DNA. T7 DNA polymerase, purified in the absence of EDTA, catalyzes only limited DNA synthesis at nicks in duplex DNA molecules, but in the presence of gene 4 protein catalyzes extensive synthesis. As synthesis proceeds, gene 4 protein facilitates unwinding of DNA with concomitant hydrolysis of a nucleoside triphosphate. Helicase activity of gene 4 protein occurs even in the absence of T7 DNA polymerase and DNA synthesis, catalyzing displacement of DNA fragments annealed to M13 single-stranded circular DNA. The helicase activity depends on dTTP and a 3'-single-stranded tail (at least six residues long) on the fragment.

When gene 4 protein reaches a specific recognition site, it catalyzes the synthesis of a tetraribonucleotide primer. On DNA templates of known sequence, the predominant recognition sequence is 3'-

32. Tabor, S., and Richardson, C. C. (1981) *PNAS 78*, 205.

CTGG-5' or 3'-CTGGT-5'. Transcription starts opposite T and RNA primers with the sequence pppACCC or pppACCA are synthesized;[33] the sequences, 3'-CTGGA/C-5' and 3'-CTGTN-5', are also used for primer synthesis, but far less frequently.

Primer synthesis in vivo and in vitro is reassuringly similar. The 5'-terminal sequences of the tetra- and pentanucleotides linked to nascent DNA pieces isolated from T7-infected cells[34] are mainly pApCpA/CpA/C. The complementary initiation site most frequently used as a signal for synthesis of primer RNA in discontinuous DNA synthesis[35] is 3'-CTGG/T. Several sites of transition from primer RNA to DNA can be observed within a short stretch of DNA, indicating that although primer RNA synthesis starts at a precisely defined nucleotide, transition to DNA synthesis can vary within two nucleotides. On a statistical basis, it may be calculated that the initiation sequence should appear about eight times per 100 nucleotides of T7 DNA and thus afford considerable flexibility for sites of RNA priming.

*Phage T4 gene 41 and gene 61 proteins.*[36,37]   This pair of proteins catalyzes the synthesis of a group of pentaribonucleotides on any natural single-stranded DNA. These oligonucleotides prime the starts of DNA chains by the seven-protein, T4 in vitro replication system (Section S5-5). Gene 41 protein alone exhibits a DNA-dependent GTPase (and ATPase) activity and in conjunction with gene 61 protein is responsible for primer synthesis and helicase activity at a replication fork. Although gene 61 protein alone has not been observed to synthesize primers, its contribution to the 41-61 function is plausibly that of a primase. Thus the paired effort of proteins 41 and 61 resembles the general (nonspecific) priming reaction carried out by the dnaB protein-primase team (Section S11-9). The mobility and helicase activity of 41-61 is reminiscent of protein n' activity in the *E. coli* primosome (Section S11-9). In comparing priming mechanisms, it is impressive that helicase and primase functions are compressed into a 60-kdal gene 4 protein by T7, are shared by a 100-kdal pair of proteins in T4, and require seven proteins (comprising 20 subunits near one million in molecular weight) for *E. coli*.

The pentaribonucleotide primers synthesized in vitro on a T4 DNA template start with a pppApC dinucleotide, followed by many differ-

33. Scherzinger, E., Lanka, E., Morelli, G., Seiffert, D., and Yuki, A. (1977) *EJB 72*, 543; Romano, L. J., and Richardson, C. C. (1979) *JBC 254*, 10476; 10483; Tabor, S., and Richardson, C. C. (1981) *PNAS 78*, 205.
34. Seki, T., and Okazaki, T. (1979) *N. A. Res. 7*, 1603; Ogawa, T., and Okazaki, T. (1979) *N. A. Res. 7*, 1621.
35. Fujiyama, A., Kohara, Y., and Okazaki, T. (1981) *PNAS 78*, 903.
36. Liu, C.-C., and Alberts, B. M. (1981) *JBC 256*, 2813; 2821.
37. Silver, L. L., Venkatesan, M., and Nossal, N. G. (1980) *ICN–UCLA Symposia 19*, 475; Nossal, N. G. (1980) *JBC 255*, 2176.

ent sequences. Similarly, the nascent DNA fragments isolated from T4-infected cells[38] have one to five ribonucleotide residues at their 5′ ends with the sequence $pApC(pN)_3$.

*Plasmid primase*. Certain conjugative-type plasmids in *E. coli* specify the synthesis of primases that can substitute for the host primase and also serve the plasmid in its maintenance and conjugative transfer. A primase of 118 kdal encoded by RP4 is needed for priming the transferred strand and for efficient priming in the discontinuous phase of plasmid replication despite the presence of the host priming systems.[39]

*Eukaryotic primase*. The nonspecificity of the decanucleotide primer (initiator RNA)[40] that precedes the nascent DNA fragments of papovaviral (p. 579) or lymphocytic DNA implies the action of a primase that counts or measures but does not read. Discovery of potent primase activity in DNA polymerase α of *Drosophila* (Section S6-2) that makes a decaribonucleotide primer has finally located the principal eukaryotic primase. A similar enzymatic activity has also been obtained from human lymphocytes.[40a] The circular, single-stranded DNA of only 1760 residues in a tiny porcine virus[41] should provide a substrate, as M13 and φX174 phages did, in probing for cellular priming mechanisms.

## S11-11 Duplex Replication: De Novo and Covalent Starts

De novo starts, as defined here, depend on covalent extension of an RNA transcript; covalent starts are direct extensions of the chromosome. A novel mechanism of initiation entails covalent attachment of the 5′ starting deoxynucleotide to a protein complexed with each end of a linear duplex genome and has been established as an early event in the replication of adenovirus (Section S15-4) and phage φ29 (Section S14-8).

### De Novo Starts

Priming by RNA polymerase initiates replication on supercoiled ColE1 plasmid, but only after a process more complex than had been anticipated (Section S14-9). The role of RNA polymerase in the

38. Kurosawa, Y., and Okazaki, T. (1979) *JMB 135*, 841.
39. Lanka, E., and Barth, P. T. (1981) *J. Bact. 148*, 769.
40. Reichard, P., and Eliasson, R. (1978) *CSHS 43*, 271.
40a. Tseng, B. Y., and Ahlem, C. N. (1982) *JBC*, in press.
41. Tischer, I., Gelderblom, H., Vettermann, W., and Koch, M.A. (1982) *Nat. 295*, 64.

initiation at the origin of the *E. coli* chromosome (oriC plasmid) (Section S11-15) or phage λ (Section S14-7) is not yet known. Nor is it clear whether T7 RNA polymerase starts replication of the T7 linear duplex by priming or by an indirect transcriptional activation (Section S14-5).

### Covalent Starts

φX174 viral strands are produced by a *rolling-circle mechanism* from a supercoiled RF, isolated protein-free, from infected cells (Fig. 11-26, p. 401). In this RF→SS reaction, the synthetic RF, isolated from a reaction mixture with the primosome nearly intact, is a superior substrate for gene A protein cleavage despite its relatively low superhelix density (Fig. S11-8). The lag in the reaction is eliminated and viral strands are produced rapidly and efficiently. Alterations in the origin sequence for gene A protein cleavage have indicated a distinctive recognition site for binding the protein that relates it to origins and promoters on other genomes (Section S13-5).

The capacity of this RF→SS system to synthesize twenty or more single-stranded circles per RF in a few minutes as the sole product has been exploited for studies of fidelity of replication (Section S11-12). It also produces from recombinant DNAs containing the gene A cleavage site highly radioactive single strands that can be used as probes in hybridization.

RF multiplication can be reconstituted by coupling the enzyme systems responsible for discontinuous complementary strand synthesis (SS→RF)[42] and continuous viral strand synthesis (RF→SS). Thirteen nearly homogeneous proteins, comprising more than thirty subunits are required.[43] Net RF synthesis proceeds in a single reaction mixture starting with viral strands (SS→RF→RF) or with either synthetic RF or the RF produced in vivo (RF→RF).[44] It is not known whether, in the RF→RF reaction, viral circles are completed before serving as templates for RF formation. The rolling-circle mechanism produces many copies of duplex DNA in a stage of phage λ replication (Fig. 14-38, p. 547), and its employment in gene amplification processes may prove to be widespread in nature (Section S13-6).

A *rolling-hairpin mechanism* in which terminal palindromic sequences are directly extended applies to the simple parvovirus genomes (Section S15-3). Whether the large eukaryotic genomes with

42. Shlomai, J., Polder, L., Arai, K., and Kornberg, A. (1981) *JBC 256*, 5233.
43. Arai, K., Arai, N., Shlomai, J., and Kornberg, A. (1980) *PNAS 77*, 3322; Arai, N., Polder, L., Arai, K., and Kornberg, A. (1981) *JBC 256*, 5239.
44. Arai, N., and Kornberg, A. (1981) *JBC 256*, 5294.

abundant palindromes in terminal (telomeric) locations also use them in replication remains to be seen.

S119

SECTION S11-12:
Fidelity of Replication

## S11-12  Fidelity of Replication

A more accurate kinetic analysis of the fidelity of replication is now available.[45] The genetic quality of phage $\phi$X174 DNA synthesized by pure proteins in the RF→SS reaction (Section S11-11)[46] is assayed by reversion of mutant phage to wild type.[47] Errors in nucleotide selection by *E. coli* DNA polymerase III holoenzyme, as detected by this approach, are largely responsible for the rate of spontaneous mutation of phage known to occur in vivo. The low frequency of spontaneous reversions, about $5 \times 10^{-7}$, is reduced a further 20-fold in replication of duplex DNA in vivo by postreplicative (mismatch) repair.[48]

Overall purine:purine misincorporation frequencies in the $\phi$X174 system in vitro are about equal to purine:pyrimidine mismatches and much greater than pyrimidine:pyrimidine mispairings,[49] thus conforming to earlier theoretical predictions.[50] The error frequency depends not only upon the base change required but also upon the neighboring DNA sequences. Phage T4 DNA polymerase as part of a reconstituted multienzyme complex (Section S14-6) shows the same properties and high degree of base-pairing fidelity observed with the host polymerase and also the mutagenic effects of substituting $Mn^{2+}$ for $Mg^{2+}$.[51]

Recent measurements of DNA polymerase I replication[52] show the high fidelity displayed by the polymerase III holoenzyme and T4 polymerase systems, in keeping with possession of the $3' \rightarrow 5'$ exonuclease editing capacity. The latter function proves to be less efficient when applied to a synthetic template, such as poly d(AT).[53] The presence of single-strand binding protein increases fidelity several fold for a variety of polymerases.[54]

Mutator effects of lowered fidelity of replication may prove to be the basis for disease and associated with aging (Section S16-3).

45. Fersht, A. R. (1981) *Proc. R. Soc. Lond. B212*, 351.
46. Fersht, A. R. (1979) *PNAS 76*, 4946.
47. Weymouth, L. A., and Loeb, L. A. (1978) *PNAS 75*, 1924.
48. Glickman, B. W., and Radman, M. (1980) *PNAS 77*, 1063.
49. Fersht, A. R., and Knill-Jones, J. W. (1981) *PNAS 78*, 4251.
50. Topal, M. D., and Fresco, J. R. (1976) *Nat. 263*, 285.
51. Hibner, U., and Alberts, B. M. (1980) *Nat. 285*, 300; Sinha, N. K., and Haimes, M. D. (1981) *JBC 256*, 10671.
52. Kunkel, T. A., and Loeb, L. A. (1980) *JBC 255*, 9961.
53. Loeb, L. A., Dube, D. K., Beckman, R. A., Koplitz, M., and Gopinathan, K. P. (1981) *JBC 256*, 3978.
54. Kunkel, T. A., Meyer, R. R., and Loeb, L. A. (1979) *PNAS 76*, 6331.

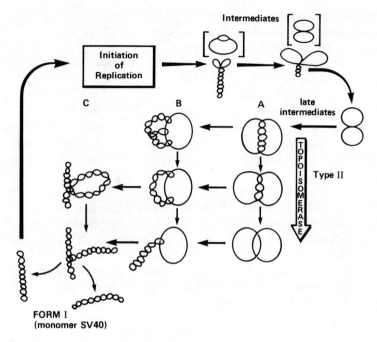

FIGURE S11-10
Proposed scheme for SV40 DNA replication and segregation. Replication
initiates on monomer SV40 DNA, then replicating structures grow
progressively larger through bidirectional fork movement reaching a point
at which 100–200 bp of unreplicated DNA remain in catenated dimers
composed of two open circles. These are gradually topoisomerized
(vertical arrows) by a type II activity to less intertwined molecules.
Independently, DNA synthesis (horizontal arrows) fills in and seals gaps,
converting form A catenated dimers to form B (one open circle interlocked
with one supercoiled circle) and then to form C molecules (two interlocked
supercoiled circles). A final topoisomerase-catalyzed step forms two
separate supercoiled monomers. [Adapted from Sundin, O., and
Varshavsky, A. (1981) *Cell 25*, 659.]

## S11-13   Termination of Replication

### Terminus in *E. coli*

A physical map of the region in which termination of the bidirectional replication of *E. coli* takes place fixes the terminus (*tre* or *terC*)
more exactly at 31.2 (±0.2) minutes,[55] near the center of the largest
genetically silent region of the chromosome.[56] This 270-kb region
contains only three genes, all nonessential, and at its margins, two of

55. Bouché, J. P. (1982) *JMB 154*, 1; Bouché, J. P., Gélugne, J. P., Louarn, J., Louarn, J. M., and
    Kaiser, K. (1982) *JMB 154*, 21.
56. Bachmann, B. J., and Low, K. B. (1980) *Microbiol. Rev. 44*, 1.

the three defective lambdoid elements in *E. coli* K12. The basis for termination at this location may not be a particular sequence, but may rather relate to its position opposite the origin (*oriC*),[57] possibly involving the terminus as part of a novel structure that embraces both replicating forks. Observed associations of the origin and terminus regions of the chromosome with membrane fractions (p. 455) may be pertinent here, but the chemical and cytologic properties of such complexes are still sadly lacking.

### Termination Mechanisms

Cutting long concatemers into genome-length units appears to account in principle for packaging a linear duplex in a phage head (p. 408). However, termination of circular DNA replication and separation of the progeny present a topological problem for the circular genomes of plasmids and many viruses and cellular chromosomes. Analysis of the segregation of replicated SV40 DNA into mature supercoiled forms may have general significance (Fig. S11-10).[58] Bidirectional replication around the 5.2-kb circle pauses 100–200 bp short of completion. DNA synthesis that continues to fill in the gaps of the "Siamese-dimer," daughter circles seals one and then the other to give two supercoiled circles held together simply by topological linkage. Topoisomerase II activity (Section S9-12) is inferred as the means of reducing the concatemers to monomeric supercoils. Interruption of segregation by placing the infected cells in hypertonic medium does not affect replication, suggesting that these processes are under independent controls that may be employed for some physiological function.

## S11-14    Enzymology of the Replicating Fork

The proteins that move the replicating fork and the means they use to do it have come into clearer view (Fig. S11-11) and begin to justify the notion of a replisome (see below), comparable in complexity and analogous in function to the ribosome.

### Replication Proteins of *E. coli*

The list of fifteen replication proteins (comprising over thirty polypeptides) (Table S11-2) has not been expanded much in the past two

---

57. Bouché, J. P., personal communication.
58. Sundin, O., and Varshavsky, A. (1980) *Cell 21*, 103; (1981) *Cell 25*, 659.

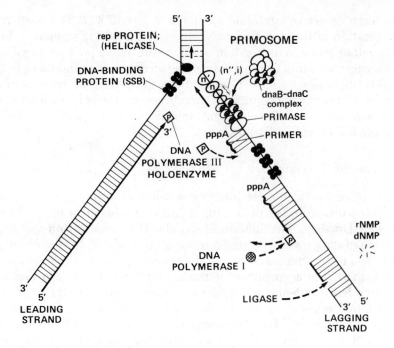

FIGURE S11-11
Proposed scheme for DNA chain growth at one of the forks of a bidirectionally replicating *E. coli* chromosome.

years although additional primosomal components (Section S11-10), helicases (Section S9-9), and holoenzyme subunits (Section S5-3) may well be discovered. Perhaps the most significant advance (compare with Table 11-5, p. 412) is that twice as many of the scarce replication proteins can now be overproduced and have thus become attractive subjects for enzymological studies.

Progress at the Fork

Opening the duplex in advance of replication would result from the action of helicases, such as rep protein acting on the 3′→5′ strand and helicase II on the 5′→3′ strand, the latter action augmented or replaced by helicase activity of the primosome. The bared single strands are promptly covered by binding protein. Synthesis by DNA polymerase III holoenzyme on the 3′→5′ strand template is continuous and matched by discontinuous synthesis of the lagging strands (Section S11-3).

TABLE S11-2
Replication proteins of *E. coli*

| Protein | Native mass, kdal | Subunits | Function | Estimated molecules/ cell | Unamp. yield, mg/kg[a] | Amplification[b] |
|---------|------------------|----------|----------|---------------------------|------------------------|------------------|
| SSB | 74 | 4 | single-strand binding | 300 | 20 | 100 |
| protein i | 66 | 3 ⎫ | | 50 | 0.5 | — |
| protein n | 28 | 2 ⎪ | primosome assembly and function | 80 | 0.4 | — |
| protein n' | 76 | 1 ⎪ | | 70 | 0.3 | — |
| protein n'' | 17 | 1 ⎬[c] | | — | — | — |
| dnaC | 29 | 1 ⎪ | | 100 | 0.2 | 50 |
| dnaB | 300 | 6 ⎪ | | 20 | 0.3 | 200 |
| primase | 60 | 1 ⎭ | primer synthesis | 50 | 0.2 | 50 |
| pol III holoenzyme | (760) | (2) | | 20 | 0.5 | — |
| α | 140 | 1 ⎫ | | — | — | — |
| ε | 25 | 1 ⎪ | processive chain elongation | — | — | — |
| θ | 10 | 1 ⎪ | | — | — | — |
| β | 37 | 1 ⎬ × 2 | | β 300[d] | 0.3 | 5 |
| γ | 52 | 1 ⎪ | | γ  20 | — | 70 |
| δ | 32 | 1 ⎪ | | — | — | — |
| τ | 83 | 1 ⎭ | | — | — | — |
| pol I | 102 | 1 | gap filling, primer excision | 300 | 10 | 70 |
| ligase | 74 | 1 | ligation | 300 | 10 | 500 |
| gyrase | 400 | 4 | supercoiling | — | — | — |
| gyrA | 210 | 2 | | 250 | 40 | 50 |
| gyrB | 190 | 2 | | 25 | 4 | 30 |
| rep | 65 | 1 | helicase | 50 | 0.6 | 10 |
| helicase II | 75 | 1 | helicase | 5000 | 1.8 | — |
| dnaA | 48 | | origin of replication | 200 | — | 50 |

[a]Final yield of purified protein:mg/kg wet weight of normal cells
[b]Normal cell protein level increased this many times by introducing plasmid or phage containing the gene.
[c]The stoichiometry of these polypeptides in the primosome is not known; the current estimate is one of each except for six each of dnaC and dnaB protomers.
[d]As dimers of β subunit

## Concurrent Replication of Both Strands

A proposal[59] that *E. coli* DNA polymerase III holoenzyme functions as a pair of identical oligomers, each with an active site, has several attractive features. Replication of both strands simultaneously by one polymerase molecule may avoid the need for repeated, time-

59. Kornberg, A., Burgers, P. M. J., and Stayton, M., unpublished observation.

consuming activations of the enzyme (Section S5-3), and by physically and functionally linking the polymerase to the primosome in a replisome structure, would coordinate their actions. Fragmentary evidence favors a dimeric polymerase: (i) a stoichiometry of two $\beta$ subunits, two $\gamma$ subunits, and two ATPs per replicating center (Section S5-3); (ii) an anomalously large size of pol III′ (Section S5-3) suggestive of a dimeric pol III core and $\tau$ subunits; (iii) looped DNA structures observed in association with the primosome (Fig. S11-7); looping of DNA is required for concurrent replication (see below) and is topologically feasible; and (iv) likelihood of two active sites in the eukaryotic DNA polymerase $\alpha$ (Section S6-2).

In the scheme for concurrent replication (Fig. S11-12), the leading strand is always ahead by the length of one nascent fragment; the regions of the parental template strands undergoing simultaneous synthesis are therefore not complementary (Fig. S11-12, stage I). By looping the lagging strand template 180° (perhaps halfway around the polymerase), the strand achieves the same 3′→5′ orientation at the fork as the leading strand (Fig. S11-12, stage II). A primer generated by the primosome is extended by polymerase as the lagging strand template is drawn through it. When synthesis approaches the 5′ end of the previous nascent fragment, the lagging strand template is released and becomes unlooped (Fig. S11-12, stage III). Synthesis of the leading strand has in the meanwhile generated a fresh, unmatched length of template for synthesis of the next nascent fragment for the lagging strand. By this mechanism, one polymerase molecule not only copies both templates concurrently but also remains linked to the primosome whose movement is in the direction *opposite* the elongation of the lagging strand.

### Replisome

This hypothetical structure, operating at each replicating fork of an *E. coli* chromosome (Fig. S11-12), might contain a dimeric polymerase, a primosome, and one or more helicases. An even more complex replisome could be envisioned should both forks of the bidirectionally replicating chromosome be contiguous and be attached to the cell membrane by specific proteins.

## S11-15  Enzymology of the Replication Origin of the *E. coli* Chromosome

"Still unanswered are the major questions of control and regulation of replication in bacterial and animal cells. What starts a cycle and what stops it? Why do embryonic cells grow and why are adult cells

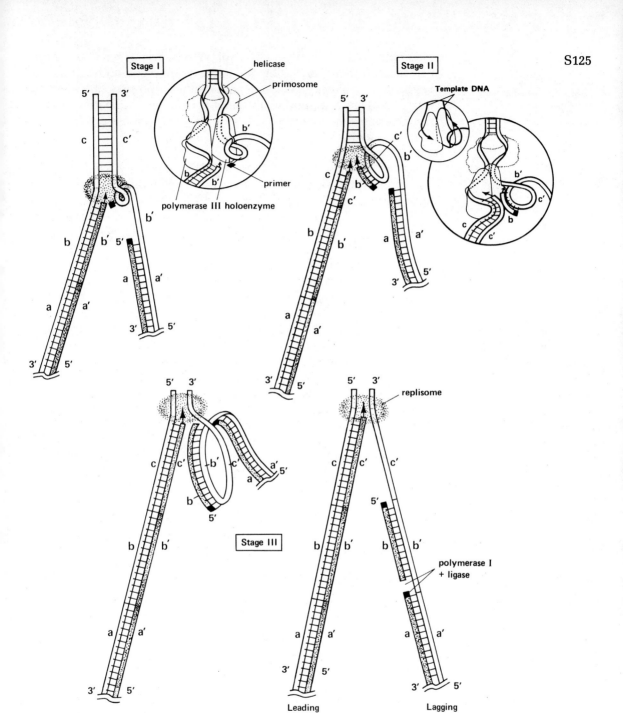

FIGURE S11-12
Proposed scheme for concurrent replication of leading and lagging strands by a dimeric polymerase associated with the primosome and one or more helicases in a "replisome." See text for details.

quiescent? What alteration of control transforms normal cells to unrestrained proliferation?" The challenge in these concluding lines of Chapter 11 (p. 413) can now be faced more optimistically with a soluble enzyme system that initiates a cycle of replication of the *E. coli* chromosome.[60]

OriC Enzyme System[60]

The availability of the chromosomal origin of *E. coli* (*oriC*) in a functional form as part of a small plasmid (p. 562; Section S13-5) or a small phage chimera was crucial in the discovery of the oriC enzyme system. An ammonium sulfate fraction prepared from cell lysates that can replicate such oriC plasmids or phage chimeras has many physiologically relevant features. This soluble enzyme system

(i)   depends completely on low levels of exogenously furnished, supercoiled oriC plasmids or phage replicative forms (RFs),

(ii)   uses only those plasmids or RFs that contain the intact *oriC* region of about 245 base pairs (Section S13-5),

(iii)   initiates replication within or near the *oriC* sequence and proceeds bidirectionally (Fig. S13-2, S13-3),[61]

(iv)   continues replication linearly, after a 5-minute lag, for 30 or more minutes to replicate completely the input DNA,[62]

(v)   depends on RNA polymerase and gyrase as indicated by total inhibition by rifampicin or nalidixate,

(vi)   depends on certain replication proteins (e.g., dnaB protein, dnaC protein, and single-strand binding protein) but not others (e.g., DNA polymerase I, recA protein), as judged by specific antibody inhibitions,

(vii)   operates independently from protein synthesis, and

(viii)   depends on dnaA activity, as shown by the inactivity of enzyme fractions from each of two $dnaA_{ts}$ mutant strains, and complementation by an enzyme fraction from a strain, containing the *dnaA* gene cloned in a plasmid with a strong promoter, that overproduces the complementing activity 50-fold, and by a nearly homogeneous dnaA protein.[63]

*Distinctive features of the oriC system.*   The reasons that made this potent system for recognition of the *E. coli* chromosome origin elusive for so many years will become apparent only when the many

60. Fuller, R. S., Kaguni, J. M., and Kornberg, A. (1981) *PNAS 78*, 7370.
61. Kaguni, J. M., Fuller, R. S., and Kornberg, A. (1982) *Nat. 296*, 623; Fuller, R. S., Kaguni, J. M., and Kornberg, A. (1981) *PNAS 78*, 7370.
62. Fuller, R. S., Kaguni, J. M., and Kornberg, A., unpublished results.
63. Fuller, R. S., Kaguni, J. M., and Kornberg, A., unpublished results.

factors that activate and inhibit it are identified and characterized. At this juncture several features are noteworthy.

S127

SECTION S11-15:
Enzymology of the
Replication Origin of
the E. coli Chromosome

(i)   Only a rather narrow ammonium sulfate fraction (Fraction II) derived from the cell lysate is active; inhibitory factors are precipitated even at modest levels of additional salt. The cell lysate itself is inactive.

(ii)   The optimal concentration of the supercoiled oriC plasmid or RF falls within a narrow range around 1 nM.

(iii)   The optimal $Mg^{2+}$ concentration lies in a very narrow range around 10 mM.

(iv)   A strong ATP-regenerating system is essential to maintain ATP at 2mM.

(v)   A flexible, hydrophilic polymer is absolutely required. Any one of several polymers, such as polyethlene glycol, polyvinyl alcohol, or methylcellulose, with molecular weights ranging from 6000 to greater than 100,000 and at concentrations of 6 to 8 percent, can serve. The polymers may increase the effective concentrations of macromolecular reactants by an "excluded volume" effect.[64]

(vi)   Contrary to expectations, the dnaA protein level is not sharply rate-limiting; the rate with Fraction II obtained from a wild-type cell is barely doubled by addition of large amounts of purified dnaA protein.

(vii)   Membranes or particulate elements appear to be absent.

*Absence of membrane involvement.*   Segregation of chromosomes should plausibly be connected with the cellular membranes that are ultimately responsible for dividing the daughter cells. Many indications of attachment of the chromosome origins to membranes (Section S13-4) include a recent observation of initiation of chromosomal DNA synthesis in crude lysates on cellophane discs.[65] Despite numerous suggestions of membrane involvement, the specific replication of oriC plasmids can now be analyzed in a soluble system. Even though membrane attachment of the origin may be a vital cellular feature, biochemical analysis of the pathway of initiation, as with fatty acid biosynthesis and certain other membrane-related processes, can proceed unimpeded by the complexities introduced by the cellular envelope.

*Purification of the oriC system.*   Isolation of the factors needed in the reaction can now be pursued with assays based on complementation of fractions from mutant extracts and by classic enzyme fractionation techniques of resolution and reconstitution (p. 381). Based

---

64. Tanford, C. (1961) *Physical Chemistry of Macromolecules,* John Wiley, New York.
65. Projan, S. J., and Wechsler, J. A. (1981) *MGG* 182, 263.

on complementation of *dnaA* mutant extracts, dnaA protein has been purified to near homogeneity.[66] The physical properties of the 48-kdal polypeptide, obtained from overproducing cells that harbor a plasmid with the *dnaA* gene, do not distinguish it qualitatively from other replication proteins. Complementation of extracts of *dnaK* mutants,[67] also deficient in *oriC* replication, provides the first opportunity to determine whether the product of this gene has an intimate role in this enzyme system.[68] With mutants in *dnaI*, regarded as a chromosome initiation gene,[69] similar studies[70] disclosed that strains from several sources are not distinguishable from *dnaA* strains, and plasmids or phages containing the *dnaA* gene suppress the *dnaI* defect.[71] A system reconstituted from the nine purified proteins needed for conversion of $\phi$X174 single strands to the duplex form (Fig. S11-1) plus RNA polymerase, gyrase, and dnaA protein is inert for *oriC* replication but can be made active with one or more factors in a crude enzyme fraction and thereby provides the assay for their purification.[72]

Progress in characterizing this new group of replication proteins should advance the understanding of how initiation of a cycle of chromosome replication is controlled (Section S13-5) and reveal more about the mechanisms of priming and fork movement in the *E. coli* chromosome.

---

66. Fuller, R. S., Kaguni, J. M., and Kornberg, A., unpublished results.
67. Georgopoulos, C. P., Lam, B., Lundquist-Heil, A., Rudolph, C. F., Yochem, J., and Feiss, M. (1979) *MGG 172*, 143; Itikawa, H., and Ryu, J. (1979) *J. Bact. 138*, 339.
68. Low, R., Kaguni, J. M., Fuller, R. S., and Kornberg, A., unpublished results.
69. Beyersmann, D., Messer, W., and Schlicht, M. (1974) *J. Bact. 118*, 783.
70. Reichard, P., Kaguni, J. M., Fuller, R. S., Bertsch, L., and Kornberg, A., unpublished results.
71. Maurer, R., personal communication; Kodaira, M., personal communication.
72. Kaguni, J. M., Bertsch, L., and Kornberg, A., unpublished results.

# Inhibitors
# of Replication

## Abstract

Inhibitors of DNA replication continue to serve the clinician as prime drugs for suppressing proliferative diseases and the laboratory investigator with the means to analyze biochemical pathways in vivo and in vitro. To the well-known classes of inhibitors that affect nucleotide biosynthesis, replication, or DNA itself, are now added agents that act on regulatory processes, including postreplicational methylation. 5-Azacytidine and related analogs have come into prominence for their suppression of mammalian DNA methylation with reverberating effects that likely include gene activation and cellular differentiation.

With more refined biochemical and structural analysis of inhibitor-oligonucleotide and inhibitor-enzyme complexes, chemical rationale is supplementing systematic screening of natural products for the discovery of superior inhibitors. The bifunctional intercalators, for example, rival polymerases and regulatory proteins in their affinity for DNA.

Many of the inhibitors considered in this chapter have mutagenic and carcinogenic effects. These agents along with others that are not included, such as polynuclear aromatic hydrocarbons (e.g., benzo[a]pyrene, dimethylbenz[a]anthracene, N-acetoxy-acetylaminofluorene) and phorbol esters (Section S13-6), would constitute a separate chapter in a new edition of *DNA Replication*.

Ionizing radiation damage to DNA has not been reviewed in this Supplement, despite its enormous importance. A National Research Council committee surveying federal programs of research on biological effects of ionizing radiation (National Academy Press, 1981; publication pending) concluded: "Despite extensive accumulation of information, . . . which may exceed the amount of knowledge that we have concerning any other noxious environmental agent, we still do not understand, nor have the data to evaluate, the basic mechanisms of radiation-induced [biological] damage. . . ." By contrast, ultraviolet-light lesions can be located with more precision and mechanisms of their repair defined in greater detail (Sections S10-8, S16-2).

## S12-1   Classes of Inhibitors

Recent advances have taken the direction of understanding mechanisms of known inhibitors rather than discovering more of them, and we have become aware of classes of inhibitors that act on regulatory processes and affect DNA replication only indirectly.[1] As examples:

(i)   Agents such as *dimethylsulfoxide, butyrate,* and *hypoxanthine* induce hemoglobin synthesis in erythroleukemic cells and the differentiation of cultured human myeloid leukemic cells into mature granulocytes. Other drugs with effects on development include *bromodeoxyuridine* or *iododeoxyuridine, steroid hormones,* and the *cytidine analogs* that influence methylation patterns (Sections S12-5, S16-7).[2]

(ii)   Although DNA is clearly the target of *anthracyclines* and *electrophilic* and *alkylating* agents (Section 12-4), their toxic effects may result from interactions with key proteins.[3]

(iii)   Factors that influence the cellular environment, such as the

1. Lotem, J., and Sachs, L. (1979) *PNAS 76,* 5158; Collins, S. J., Bodner, A., Ting, R., and Gallo, R. C. (1980) *Int. J. Cancer 25,* 213.
2. Razin, A., and Riggs, A. D. (1980) *Science 210,* 604.
3. Harrap, K. R., Jeney, A., Thraves, P. J., Riches, P. J., and Wilkinson, R. (1981) in *Molecular Actions and Targets for Cancer Chemotherapeutic Agents* (Sartorelli, A. C., Lazo, T. J., and Bertino, J. R., eds.) Academic Press, New York, p. 45.

levels of oxygen[4] and free-radical scavengers,[5] can have decisive effects on drug actions. *Bleomycins* and *neocarzinostatin* depend on oxygen; *anthracycline*, *hydroxyellipticine*, and *hydroxyurea* actions involve free radicals.

(iv)   Chemical rationale rather than systematic screening of natural products has produced inhibitors that mimic transition-state intermediates in key enzymatic reactions (e.g., PALA in Section 12-2, p. 420) or perform as superior complexing agents (e.g., bifunctional intercalators in Section S12-4; methotrexate with dihydrofolate reductase[6]). X-ray diffraction and NMR analyses[7] of oligonucleotide drug complexes have helped in visualizing intercalative and other interactions.

(v)   The crucial dependence of all cells on DNA repair mechanisms (Sections S16-2, S16-3) makes these systems key targets for virtually all agents affecting DNA and its replication.[8] An example is the alkylating agent, *bis-chloroethylnitrosourea* (Section 12-4, p. 431), which acts by transfer of a chloroethyl group to oxygen-6 of guanine, which in turn provides an electrophilic center for crosslinking duplex DNA.[9] Mutants unable to excise the $O^6$-alkyl guanine are far more sensitive to the cytotoxic action of the drug.[10] The frequent occurrence of similar phenotypes in human cancer cells and SV40-transformed cells may explain the preferential chemotherapeutic efficiency of the nitrosoureas. Cytotoxicity of *cis-diamminedichloroplatinum (II)* in *E. coli* is also influenced by the repair proficiency of the strain.[11]

## S12-2   Inhibitors of Nucleotide Biosynthesis

Several compounds, in addition to those considered in Section 12-2 (p. 418), deserve attention.

### Purine Synthesis

*Hadacidin* (N-formyl-N-hydroxyglycine) (Fig. S12-1) is an aspartic acid analog isolated from fungi and inhibitory to growth of bacteria, tumor cells, and plant tissues. It competes with aspartic acid to

4. Teicher, B. A., Lazo, J. S., and Sartorelli, A. C. (1981) *Canc. Res. 41*, 73.
5. Harman, D. (1981) *PNAS 78*, 7124.
6. Quigley, G. J., Wang, A. H.-J., Ughetto, G., van der Marel, G., van Boom, J. H., and Rich, A. (1980) *PNAS 77*, 7204.
7. Sakore, T. D., Reddy, B. S., and Sobell, H. M. (1979) *JMB 135*, 763; 813; Patel, D. J., Kozlowski, S. A., Rice, J. A., Broka, C., and Itakerra, K. (1981) *PNAS 78*, 7281.
8. Hanawalt, P. C., Cooper, P. K., Ganesan, A. K., and Smith, C. A. (1979) *ARB 48*, 783.
9. Erickson, L. C., Laurent, G., Sharkey, N. A., and Kohn, K. W. (1980) *Nat. 288*, 727.
10. Day, R. S., III, Ziolkowski, C. H. J., Scudiero, D. A., Meyer, S. A., Lubiniecki, A. S., Girardi, A. J., Galloway, S. M., and Bynum, G. D. (1980) *Nat. 288*, 724.
11. Brouwer, J., Van de Putte, P., Fichtinger-Schepman, A. M. J., and Reedijk, J. (1981) *PNAS 78*, 7010.

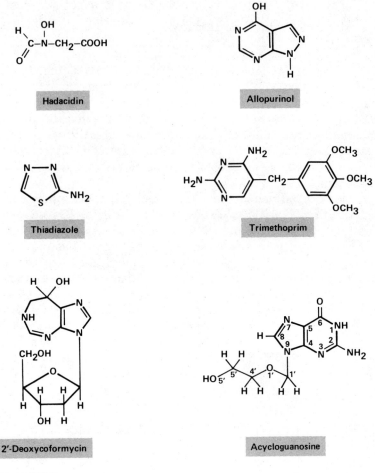

FIGURE S12-1
Inhibitors of nucleotide and DNA synthesis and catabolite analogs.

inhibit adenylosuccinate synthetase and thereby the de novo synthesis of AMP.[12] _Thiadiazole_ (2-amino-1,3,4-thiadiazole) (Fig. S12-1)[13] is, in addition to _mycophenolic acid_ and _ribavirin monophosphate,_ an inhibitor of IMP dehydrogenase and thus of guanylate biosynthesis.

Pyrimidine Synthesis

_5-Azacytidine_ (Fig. S12-4) inhibits macromolecular biosyntheses and functions through (i) being incorporated into RNA and DNA (Section S12-3), (ii) inhibiting DNA methylation (Section S12-5), and (iii) com-

12. Shigeura, H. T., and Gordon, C. N. (1962) _JBC 237,_ 1932; 1937; Rossomando, E. F., Maldonado, B., and Crean, E. V. (1978) _Antimicrobial Agents and Chemotherapy 14,_ 476.
13. Nelson, J. A., Rose, L. M., and Bennett, L. L., Jr. (1977) _Canc. Res. 37,_ 182.

peting at the nucleotide level, as does *6-azauridine* (Table 12-1, p. 419), with the decarboxylation of orotidine monophosphate to uridylate.

## Folate Synthesis

*Trimethoprim* (Fig. S12-1), an inhibitor of dihydrofolate (folate) reductase, fulfills the same role as *methotrexate* (Table 12-1, p. 419). The use of methotrexate and *5-fluorouracil* in combination for the treatment of cancer may result in antagonism as well as synergism between these two drugs.[14] Inhibition of thymidylate synthesis by 5-fluorodeoxyuridylate (FdUMP) requires methylene tetrahydrofolate for covalent binding to thymidylate synthetase. Because inhibition of dihydrofolate reductase by methotrexate prevents regeneration of the tetrahydrofolate, FdUMP is unable to form the ternary complex necessary for prolonged inhibition. Another damaging effect of methotrexate is the disruption of DNA due to excessive incorporation and excision of uracil resulting from high levels of dUTP relative to dTTP (Section S2-8).

## Deoxynucleotide Synthesis

*Hydroxyurea* inhibition of the mammalian ribonucleoside diphosphate reductase (Section S2-6) has helped in the analysis of the free radical mechanism of the reaction.

## Catabolite Analogs

The effectiveness of biosynthetic analogs often depends on their being converted to the nucleotide form and avoiding removal by catabolic enzymes. Potentiation of *6-mercaptopurine* antitumor and toxic activity by *allopurinol* (Fig. S12-1, Section S12-3) may be due to inhibition of xanthine oxidase by the latter. Inhibitors of adenosine deaminase not only potentiate the antitumor activity of adenosine analogs, but one of these, *deoxycoformycin* (Fig. S12-1), even when administered alone, produces remissions in patients with acute leukemia derived from T lymphocytes.[15]

## S12-3   Nucleotide Analogs Incorporated into DNA or RNA

In addition to the analogs listed in Section 12-3 (p. 423), 5-azacytidine and allopurinol should be mentioned.

   *5-Azacytidine* (Fig. S12-4) (Sections S12-2, S12-5) incorporation

14. Cadman, E., Heimer, R., and Benz, C. (1981) *JBC 256*, 1695.
15. Cass, C. E. (1979) in *Antibiotics V.* (Hahn, F. E., ed.) Springer-Verlag, New York, p. 85.

into RNA and DNA may account for defective processing of ribosomal RNA,[16] but is not demonstrably mutagenic.

*Allopurinol* (Fig. S12-1), the synthetic analog of hypoxanthine (positions of nitrogen-7 and carbon-8 reversed) is in widespread clinical use to inhibit xanthine (and hypoxanthine) oxidase and thereby suppress the accumulation of uric acid in gout and related disorders. The drug is also an antiprotozoal agent by virtue of being incorporated into RNA. Allopurinol is converted first to the ribonucleotide in mammals as well as in protozoa, but only in protozoa is it then accepted by adenylosuccinate synthetase and lyase to become the AMP analog, which as a nucleoside triphosphate is incorporated into RNA.[17]

## S12-4 Inhibitors That Bind to or Modify DNA[18]

Beyond the inhibitors already described in Section 12-4 (p. 427), some additional agents, particularly the bifunctional intercalators and the covalent modifiers, are noteworthy.

### Noncovalent DNA Binders

*Monofunctional intercalators.* *Nogalamycin* (Fig. 12-5, p. 429) and related nogalose-containing antibiotics bind preferentially to AT-rich regions. Nogalamycin prevents the de novo synthesis by *E. coli* DNA polymerase I of the alternating copolymer poly d(AT) thus enabling the homopolymer pair poly dA · poly dT to be formed. However, nogalamycin derivatives do not display this preference,[19] for reasons that are still obscure. *Ethidium bromide* has come into wide use as a fluorescent probe for detecting and measuring levels of RNA and DNA.[20] *Ellipticines* (Fig. S12-2) are plant alkaloids with antitumor action that may depend on the production of free radicals or transient electrophilic compounds.[21]

*Bifunctional intercalators.* Several classes of compounds intercalate at two DNA sites simultaneously:[22] the synthetic[23] *acridines,*

16. Weiss, J. W., and Pitot, H. C. (1975) *B.* 14, 316.
17. Nelson, D. J., Bugge, C. J. L., Elion, G. B., Berens, R. L., and Marr, J. J. (1979) *JBC 254,* 3959; Spector, T., Jones, T. E., and Elion, G. B. (1979) *JBC 254,* 8422.
18. Waring, M. J. (1981) *ARB 50,* 159; (1980) *Nucleic Acid Geometry and Dynamics* (Sarma, R. H., ed.) Pergamon Press, New York.
19. Richardson, C. L., Grant, A. D., Schpok, S. L., Krueger, W. C., and Li, L. H. (1981) *Canc. Res.* 41, 2235.
20. Le Pecq, J.-B., and Paoletti, C. (1967) *JMB 27,* 87.
21. Le Pecq, J.-B., Gosse, C., Dat-Xuong, N., and Paoletti, C. (1974) *PNAS 71,* 5078; Auclair, C., and Paoletti, C. (1981) *J. Med. Chem.* 24, 289.
22. Le Pecq, J.-B., Le Bret, M., Barbet, J., and Roques, B. (1975) *PNAS 72,* 2915.
23. Kuhlmann, K. F., and Mosher, C. W. (1981) *J. Med. Chem.* 24, 1333.

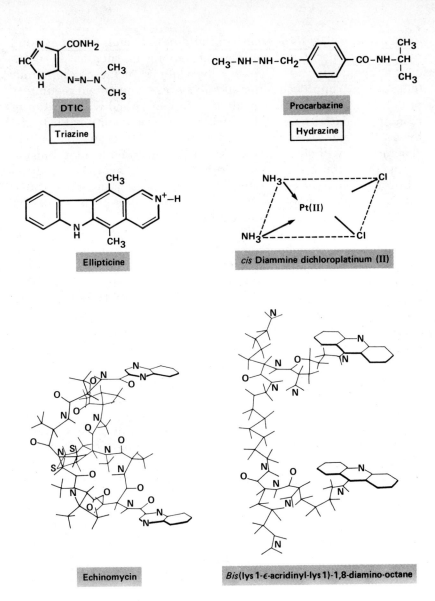

FIGURE S12-2
Inhibitors that bind DNA covalently and noncovalently.

*anthracyclines, chloroquines, quinaldines, phenanthridines,* and *pyridocarbazoles,*[24] and the natural antibiotics, such as the quinoxalin, *echinomycin.*[25] The very high affinity of these compounds (near $10^{11}M^{-1}$) approaches that of polymerases and potent regulatory proteins.[26] A curious biological effect, not observed with antitumor monointercalators (e.g., actinomycin D or adriamycin) is illustrated by a 7H-pyridocarbazole dimer. Exposed cells grow for several generations, without apparent effect on macromolecular biosynthesis, then stop dividing and do not recover. This delayed, drug-induced mortality may be related to a cellular program of controlled suicide (called "apoptosis")[27] and may be useful as a model for similar events in embryonic development.

*Echinomycin (quinomycin A)* (Fig. S12-2), a quinoxaline antibiotic produced by streptomycetes, is highly active against Gram-positive bacteria, viruses, and tumors. It binds different DNA sites with different affinities showing some preference for GC-rich regions.[28] The echinomycin structure is the basis for design of new peptidic bifunctional intercalating agents such as a *lysyl-lysine bifunctional derivative of 9-aminoacridine* (Fig. S12-2).[29]

Covalent DNA Binders (Modifiers)

*Bleomycins* (Fig. 12-7, p. 432).[30]   These antibiotics upon activation exert their cytotoxic activity by cleaving DNA at purine-pyrimidine (GpT and GpC sequences) and pyrimidine-pyrimidine sequences.[31] The activated drug is a transient complex of iron and oxygen; Fe(II) plus $O_2$ is effective and so is Fe(III) plus $H_2O_2$ (or other hydroperoxides).[32] Interaction with DNA (in vitro) is a two-step process. First, the agent intercalates through its tripeptide cationic terminal region by binding along a potential Cu(II) chelation site localized around the bleomycin pyrimidine moiety. Next, bases, mostly pyrimidines, are released by rupture of the N-glycosylic bond;

24. Pelaprat, D., Delbarre, A., Le Guen, I., Roques, B.-P., and Le Pecq, J.-B. (1980) *J. Med. Chem.* 23, 1336.
25. Waring, M. (1981) *ARB 50,* 159.
26. Capelle, N., Barbet, J., Dessen, P., Blanquet, S., Roques, B. P., and Le Pecq, J.-B. (1979) *B.* 18, 3354.
27. Wyllie, A. H. (1980) *Nat. 284,* 555.
28. Fox, K. R., Wakelin, L. P. G., and Waring, M. J. (1981) *B.* 20, 5768.
29. Bernier, J.-L., Henichart, J.-P., and Catteau, J.-P. (1981) *BJ 199,* 479.
30. Hecht, S. M. (1979) *Bleomycin: Chemical, Biochemical and Biological Aspects.* Springer-Verlag, New York.
31. Takeshita, M., Kappen, L. S., Grollman, A. P., Eisenberg, M., and Goldberg, I. H. (1981) *B.* 20, 7599.
32. Burger, R. M., Peisach, J., and Horwitz, S. B. (1981) *JBC 256,* 11636; Sakai, T. T., Riordan, J. M., Booth, T. E., and Glickson, J. D. (1981) *J. Med. Chem. 24,* 279; Kuo, M. T. (1981) *Canc. Res.* 41, 2439.

fragmentation of the labilized AP (apurinic, apyrimidinic) DNA follows. There are one-tenth as many double-strand breaks as single-strand breaks. Excision-repair at AP sites is induced in *E. coli* and yeast.

Bleomycins, in their nuclease-like actions, can preferentially destroy DNA sequences in chromatin with a transcriptionally active configuration. Some drug actions are amplified by sulfhydryl and other reducing agents and by a wide range of aromatic compounds, such as acridines, anthracenes, and triphenylmethane dyes.

*Neocarzinostatin* (p. 434). This drug and the closely related auromomycin introduce single-strand breaks in linear duplex or superhelical DNA to release free bases in an oxygen-dependent reaction greatly stimulated by mercaptans. Neocarzinostatin primarily attacks thymine (75 percent of the total), but also adenine (19 percent) and cytosine (6 percent); auromomycin attacks guanine (67 percent), thymine (24 percent), and adenine (9 percent).[33] Oxidation of the 5' carbon of nucleosides in DNA to the aldehyde results in a strand break and a DNA fragment bearing nucleoside-5'-aldehyde at its 5' end. As with bleomycin damage, the DNA with multiple AP sites is susceptible to further fragmentation or to repair.

The apoprotein (acidic polypeptide) of neocarzinostatin has no biologic activity, whereas the nonprotein chromophore possesses full activity; the apoprotein stabilizes the labile chromophore and may control its release for interaction with DNA. The chromophore ($C_{35}H_{35}NO_{12}$) contains 2,6-dideoxy-2(methylamino) galactose and 2-hydroxy-5-methoxy-7-methylnaphthoate covalently linked to a highly saturated $C_{15}H_{10}O_4$ unit that contains a 5-membered cyclic carbonate ring (1,3-dioxolan-2-one.)[34]

*Psoralens (furocoumarins)* (p. 434). Further studies exploit these intercalative compounds as reagents for analysis of nucleic acid structure and as possible therapeutic drugs for psoriasis and mycosis fungoides.[35]

*Cis-diamminedichloroplatinum (II)*[36] (Fig. S12-2). Inhibition of cell division in *E. coli* by neutral platinum coordination compounds

33. Takeshita, M., Kappen, L. S., Grollman, A. P., Eisenberg, M., and Goldberg, I. H. (1981) *B. 20*, 7599; Povirk, L. F., and Goldberg, I. H. (1982) *PNAS 79*, 369; Kappen, L. S., Goldberg, I. H., and Liesch, J. M. (1982) *PNAS 79*, 744.
34. Napier, M. A., Holmquist, B., Strydom, D. J., and Goldberg, I. H. (1981) *B. 20*, 5602.
35. Povirk, L. F., and Goldberg, I. H. (1980) *B. 19*, 4773; Tatsumi, K., Bose, K. K., Ayres, K., and Strauss, B. S. (1980) *B. 19*, 4767; Goldberg, I. H., Hatayama, T., Kappen, L. S., Napier, M. A., and Povirk, L. F. (1981) in *Molecular Actions and Targets for Cancer Chemotherapeutic Agents* (Sartorelli, A. C., Lazo, T. J., and Bertino, J. R., eds.) Academic Press, New York, p. 163.
36. Roberts, J. J., and Thomson, A. J. (1979) *Prog. N. A. Res. 22*, 71.

(e.g., *cis-Pt(NH₃)₂Cl₂*) led to trials in cancer chemotherapy. For anti-tumor activity, the compounds must have at least two labile ligands in a *cis* configuration. Covalent binding to DNA is responsible for cytotoxicity.[37] For an equal number of Pt atoms bound to DNA, *cis*-Pt(NH₃)₂Cl₂ inhibits DNA synthesis more efficiently than the *trans* isomer, cell killing is 5–10 times greater, and mutagenesis is at least several hundred fold higher.[38] In vitro, *cis* and *trans* platinum coordination complexes bind guanine primarily, most likely at nitrogen-7. Aside from crosslinks that account for only 1 percent of the lesions, little is known about the binding of platinum in vivo. Sensitivity to *cis*-Pt(NH₃)₂Cl₂ of repair-deficient *E. coli* mutants suggests that the DNA lesions are removed by both UV- and SOS-repair mechanisms.[39]

_Alkylating compounds_ or their metabolites that modify DNA are in common use in cancer chemotherapy. To be added to those listed (Fig. 12-6, p. 431) are a _triazine, DTIC_ [5-(3,3-dimethyl-1-triazine) imidazole-4-carboxamide] and a _hydrazine, procarbazine_ [1-methyl-2-p-(isopropylcarbamobenzylhydrazine)] (Fig. S12-2).

## S12-5    Inhibitors That Bind to Polymerases and Replication Proteins

DNA Polymerase Inhibitors

_Aphidicolin_ (p. 439), a tetracyclic diterpenoid antibiotic, inhibits replicative eukaryotic DNA polymerases ($\alpha$ and $\delta$ polymerases of animals, viral-encoded polymerases, and $\alpha$-like polymerases of yeast[40] and plants); it does not affect the $\beta$ and $\gamma$ polymerases (Sections S6-1, S6-2). Using aphidicolin as a selective reagent, aphidicolin-resistant DNA polymerase $\alpha$ mutants have been isolated from *Drosophila*[41] and other cultured cells.[42] Aphidicolin inhibition of DNA replication of phages $\phi$29 and M2, but not of their *B. subtilis* host or other *B. subtilis* phages,[43] suggests that aphidicolin acts at the 5'-phosphoryl-linked phage protein on which $\phi$29 (and presumably M2) appears to rely for initiation of DNA synthesis (Section

37. Butour, J.-L., and Macquet, J.-P. (1981) *BBA 653*, 305.
38. Johnson, N. P., Hoeschele, J. D., Rahn, R. O., O'Neill, J. P., and Hsie, A. W. (1980) *Canc. Res. 40*, 1463; Brouwer, J., Van de Putte, P., Fichtinger-Schepman, A. M. J., and Reedijk, J. (1981) *PNAS 78*, 7010.
39. Brouwer, J., Van de Putte, P., Fichtinger-Schepman, A. M. J., and Reedijk, J. (1981) *PNAS 78*, 7010.
40. Huberman, J. A. (1981) *Cell 23*, 647.
41. Sugino, A., and Nakayama, K. (1980) *PNAS 77*, 7049.
42. Nishimura, M., Yasuda, H., Ikegami, S., Ohashi, M., and Yamada, M. (1979) *BBRC 91*, 939.
43. Hirokawa, H., Matsumoto, K., and Ohashi, M. (1981) Poster presentation at "Molecular Cloning and Gene Regulation in Bacilli" Conference, Stanford, Calif.

S14-8). Aphidicolin is a strong competitor of dCMP incorporation, less so of dTMP, and shows little if any competition with dAMP and dGMP.[44]

S139

SECTION S12-5:
Inhibitors That Bind to
Polymerases and
Replication Proteins

*6-Anilinouracils*,[45] potent inhibitors of DNA polymerase III of *B. subtilis*, function by the mechanism previously described for the arylazopyrimidines (p. 181).

*Acycloguanosine* (*Acyclovir*) (Fig. S12-1), 9-(2-hydroxyethoxy-methyl) guanine, one of several natural and synthetic nucleoside analogs (9-(2,3-dihydroxypropyl)adenine is another),[46] is a potent inhibitor of herpes simplex viruses. The drug is effective because the analog is phosphorylated to acycloGMP by the herpes-encoded, but not by the host cell, *thymidine* kinase. AcycloGMP is then phosphorylated by host cell kinases to acycloGTP, which inhibits the herpes-encoded DNA polymerase (Section S6-7) more profoundly than it does the host DNA polymerase.[47] An x-ray diffraction analysis of the acycloguanosine structure[48] attempts to rationalize the remarkable enzyme specificities of the nucleoside analog and its phosphorylated derivatives.

## Post-Replicational Modifications

Except for some phage systems, modifications of DNA and RNA are made after the chain has been assembled. The functional significance of these modifications is becoming clearer, and with this appreciation greater interest attaches to agents that affect these processes. For example, the methylation of certain cytosine residues in mammalian DNA is likely to have an important influence on gene activation and cell differentiation (Section S16-7).[49] The methyl donor is invariably S-adenosylmethionine (AdoMet or SAM) (Fig. S12-3) and the compounds that affect (i) methyl transferases, (ii) hydrolytic enzymes that remove inhibitory products, and (iii) the enzymes of AdoMet regeneration can have major consequences. AdoMet is also the source of the propylamine moiety in the biosynthesis of spermidine and spermine, the polyamines with pervasive influence on nucleic acid functions and cellular metabolism.[50]

44. Chang, C.-C., Boezi, J. A., Warren, S. T., Sabourin, C. L. K., Liu, P. K., Glatzer, L., and Trosko, J. E. (1981) *Somat. Cell. Genet. 7*, 235; Krokan, H., Wist, E., and Krokan, R. H. (1981) *N. A. Res. 9*, 4709.
45. Wright, G., and Brown, N. (1980) *J. Med. Chem. 23*, 34.
46. De Clercq, E., and Holy, A. (1979) *J. Med. Chem. 22*, 510; King, G. S. D., and Sengier, L. (1981) *J. Chem. Res.* (M) 1501.
47. Furman, P. A., St. Clair, M. H., Fyfe, J. A., Rideout, J. L., Keller, P. M., and Elion, G. B. (1979) *J. Virol. 32*, 72; Miller, W. H., and Miller, R. L. (1980) *JBC 255*, 7204.
48. Birnbaum, G. I., Cygler, M., Kusmierek, J. T., and Shugar, D. (1981) *BBRC 103*, 968.
49. Jones, P. A., and Taylor, S. M. (1980) *Cell 20*, 85; (1981) *N. A. Res. 9*, 2933.
50. Heby, O. (1981) *Differentiation 19*, 1.

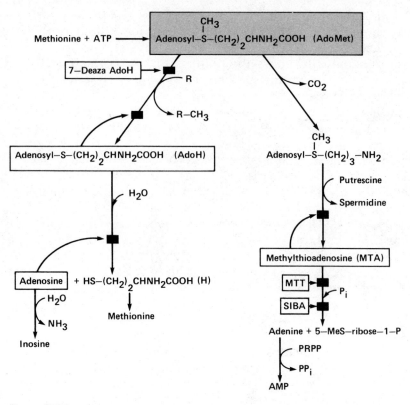

FIGURE S12-3
Pathways of S-adenosylmethionine utilization and the inhibitors that affect them.

5-Azacytidine (Sections S12-2, S12-3). This cytidine analog and others modified in the methyl acceptor position of the pyrimidine ring induce undermethylation of mammalian DNA presumably by inhibiting the methyl transferase. Three related analogs, 5-aza-2'-deoxycytidine, pseudoisocytidine, and 5-fluoro-2'-deoxycytidine (Fig. S12-4)[51] are thought to exert similar effects as do 3-deazaadenosine and ethionine (Fig. S12-4).[52] The latter, long known to be the cause of liver cancer and an antagonist of methionine, may exert its carcinogenic effect through the alteration of DNA methylation patterns (Section S16-7).

The products of methyl and propylamine transfers are, respectively, S-adenosylhomocysteine (AdoH) and 5'-deoxy-5-methylthioadenosine (MTA). These products inhibit the respective group

51. Jones, P. A., and Taylor, S. M. (1980) Cell 20, 85; (1981) N. A. Res. 9, 2933.
52. Southern, E. M. (1975) JMB 98, 503; Maniatis, T., Jeffrey, A., and Kleid, D. G. (1975) PNAS 72, 1184.

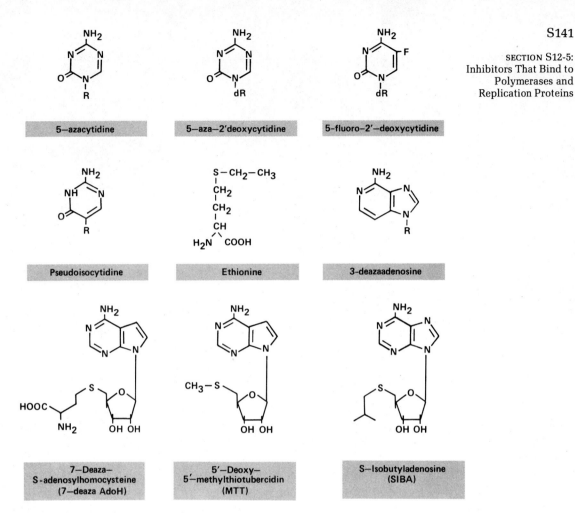

FIGURE S12-4
Inhibitors that affect nucleic acid methylation.

transfer reactions but they are actively catabolized (Fig. S12-3). Some metabolically stable, synthetic analogs of AdoH and MTA have been prepared.[53] *7-Deaza-S-adenosylhomocysteine* is a potent inhibitor of AdoMet-dependent methylases and decreases the level of methylation of tRNAs and mRNAs in animal cells. *5'-Deoxy-5-methyl-thiotubercidin* (MTT) and *S-isobutyladenosine* (SIBA) (Fig. S12-4) are inhibitors of MTA phosphorylase.

53. Parks, R. E., Stoeckler, J. D., Cambor, C., Savarese, T. M., Crabtree, G. W., and Chu, S. H. (1981) in *Molecular Actions and Targets for Cancer Chemotherapeutic Agents* (Sartorelli, A. C., Lazo, J. S., and Bertino, J. R., eds.) Academic Press, New York, p. 229; Kaehler, M., Coward, J., and Rottman, F. (1979) *N. A. Res.* 6, 1161.

# S13

# Regulation
# of Replication

## Abstract

Initiation of a cycle of chromosome replication is the prime target of its regulation. The switch that governs this process and how it is operated has not been clearly seen in studies with intact cells. Although enzymatic analysis of viral replicative cycles has been an indispensable guide to cellular mechanisms of ongoing replication, the uncontrolled nature of the multiplication of viruses disqualifies these systems as models for regulation. Major advances have therefore come from examining autonomous plasmids whose copy number is under cellular control and of chimeric plasmids engineered to depend on the very origin of the cellular chromosome to govern their replication. Determining the nucleotide sequence of the origin region and manipulating it by substitutions, deletions, insertions, and translocations have provided important clues.

Inevitably, the factors that bind DNA sequences to activate or repress their use are exposed only by resolution and reconstitution of the system in vitro. The first surprise revealed by such studies is that the control of replication of ColE1, R1, and related plasmids in *E. coli* is vested not in a protein as might have been expected but rather in a small RNA transcript of the origin region (see Section S14-9). Another promising development is the discovery of a soluble enzyme system specific for plasmids that contain the *E. coli* origin sequence (oriC). Bidirectional replication starting from that sequence depends on dnaA protein, RNA polymerase, gyrase, and the numerous replication proteins (e.g., polymerases, binding proteins, ligases) (see Section S11-15). With purification of dnaA protein in a functional form, its complex physiological behavior, implied by interactions with RNA polymerase and many other gene products, can now be analyzed.

In eukaryotic cells, the developments are not as advanced but follow similar directions. Autonomously replicating yeast plasmids are available that respond in vivo to cell-development-cycle control, and a soluble yeast system has been discovered that replicates plasmids from a unique origin and depends on many proteins, among which at least one is the product of an essential cell-development-cycle gene (see Section S15-8). In addition to the mechanisms that regulate replication in proliferating cells, an especially intriguing problem in replication is the mechanism of massive gene amplification observed in terminally differentiated cells.

The likely participation of membranes in replication in vivo invariably influences experimental approaches to regulation of replication in vitro. Nevertheless, the effective in vitro systems for plasmid replication, including that of oriC with its absolute dependence on dnaA protein, have shown no evidence of membrane involvement at any stage of the replication process.

## S13-4  Bacterial Chromosome-Membrane Association

The number and variety of observations that associate the chromosome, particularly the origin, terminus, and replicating fork, with cell-envelope membranes keep increasing.[1] Binding of a 460-bp region, containing the origin of *E. coli* replication, to outer membrane fractions is facilitated by two protein fractions.[2] Mutants in

---

1. Doyle, R. J., Streips, U. N., Imada, S., Fan, V. S. C., and Brown, W. C. (1980) *J. Bact. 144*, 957; Sandler, N., and Keynan, A. (1981) *J. Bact. 148*, 443.
2. Hendrickson, W., Yamaki, H., Murchie, J., King, M., Boyd, D., and Schaechter, M. (1981) *ICN–UCLA Symposia, 22*, 79.

lipid metabolism are inhibited in DNA replication in *Caulobacter*.[3] In *B. subtilis* the specific binding of origin DNA to membrane is regulated by mutations that control initiation of DNA replication.[4] The mutational control appears to be replicon-specific inasmuch as one mutation inhibits both the chromosome and plasmids in their replication and membrane binding, while another does not affect plasmid replication or membrane binding but does inhibit these functions for the host chromosome.

Despite these numerous indications of a connection between replication and membranes, no clear instance is known of a replicative function that depends on attachment to a membrane. On the contrary, two findings make the connection between membranes and the initiation of a cycle of replication seem less obligatory. One concerns the initiation of *B. subtilis* chromosome replication.[5] The origin DNA is found first in a nonmembrane complex and only after considerable time and replication does the membrane connection become manifest. The other finding is a vigorous and soluble enzyme system from *E. coli* that recognizes the *E. coli* chromosome origin for the start of bidirectional replication without any dependence on membranes (Section S11-15). Moreover, protein D, an outer-membrane protein of *E. coli* proposed as a unique DNA linker (p. 456), is now recognized as a family of induced iron-transport proteins,[6] and as with other outer-membrane proteins their synthesis is not coupled to cycles of DNA replication.[7]

All the evidence indicates that if replisomes and chromosomes are associated with cell membranes, they are tethered to a membrane surface by ionic and other dissociable interactions that are not essential for their enzymatic functions. Furthermore, should membrane-chromosome associations be proved, their function may be related as much to segregation at cell division as to initiation of a cycle of replication.

## S13-5 Replication Origins and Control of Copy Number

Initiation of the prokaryotic chromosome is likely to be the key locus for regulating the whole process. Insights gained from defining

3. Contreras, I., Bender, R., Weissborn, A., Amemiya, K., Mansour, J., Henry, S., and Shapiro, L. (1980) *JMB 138*, 401; *JMB*, in press.
4. Winston, S., and Sueoka, N. (1980) *PNAS 77*, 2834.
5. Yoshikawa, H., Yamaguchi, K., Seiki, M., Ogasawara, N., and Toyoda, H. (1978) *CSHS 43*, 569; Yoshikawa, H., Ogasawara, N., and Seiki, M. (1980) *MGG 179*, 265; Ogasawara, N., Seiki, M., and Yoshikawa, H. (1981) *MGG 181*, 332.
6. Boyd, A., and Holland, I. B. (1977) *FEBS Lett. 76*, 20.
7. Boyd, A., and Holland, I. B. (1979) *Cell 18*, 287; Klebba, P. (1981) Ph.D. thesis, Univ. of Calif., Berkeley.

origins of replication more precisely are the subject of this section. The factors that control copy number by regulating initiation at the origin are now being discovered in the refined in vitro studies of replication of ColE1 and related plasmids (Section S14-9). Insights into control may also be forthcoming from analysis of an enzyme system for replication of plasmids or phage chimeras directed by the *E. coli* chromosome origin (oriC) (Section S11-15).

Replication Origins

*Phage origins*. With the technical ease of synthesizing and determining DNA sequences and altering their composition and genomic location, several interesting observations have been made.

(i) *Complementary strand of single-stranded phages*. The origins for synthesis of complementary strands of filamentous phages (M13, fd) and spherical phages (G4 and $\phi$X174) are each recognized and utilized by specific host enzyme systems. Transfer to M13 of the G4 origin or the $\phi$X174 origin[8] (the protein n' recognition site at which assembly of the mobile primosome is initiated) endows M13 with a choice between alternative host systems for priming replication. Recombinants are selected on the basis that M13 is dependent on rifampicin-sensitive (RNA polymerase) priming; priming of G4 by primase and $\phi$X174 by the primosome system is rifampicin resistant.

(ii) *Viral strand of single-stranded phages*. The origin of viral strand synthesis in the duplex replicative form is recognized by a phage-encoded protein—gene A for $\phi$X174 and G4, gene 2 for M13. Cleavage of the viral strand at the origin marks the start of rolling-circle replication. The 30-bp origin region, strongly conserved among six phages related to $\phi$X174,[9] can be transferred to plasmids, which are then endowed with the same replicative capacity. The great efficiency with which single-stranded viral circles are produced in vitro has been exploited to obtain from cloned recombinant DNA highly radioactive probes in single-stranded form.[10]

A synthetic, single-stranded decameric sequence that contains the gene A recognition site (residues 4299–4308 in Fig. S13-1) is cleaved by the enzyme, as is the single strand of polyoma DNA that surprisingly also has this sequence.[11] However, this decamer linked to the adjoining decameric AT-rich sequence (residues 4299–4318) as

8. Kaguni, J., and Ray, D. S. (1979) *JMB 135*, 863; Kaguni, J., LaVerne, L. S., Strathearn, M., and Ray, D. S. (1980) in *DNA-Recombination Interactions and Repair* (Zadrazil, S., and Sponar, J., eds.) Pergamon Press, New York, p. 35.

9. Heidekamp, F., Baas, P. D., and Jansz, H. S. (1982) *J. Virol 42*, 91.

10. Zipursky, S. L., Reinberg, D., and Hurwitz, J. (1980) *PNAS 77*, 5182.

11. van Mansfeld, A. D. M., Langeveld, S. A., Baas, P. D., Jansz, H. S., van der Marel, G. A., Veeneman, G. H., and van Boom, J. H. (1980) *Nat. 288*, 561.

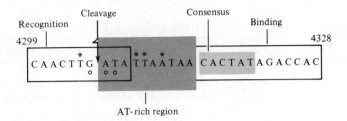

FIGURE S13-1
Sequence of the origin of φX174 DNA for viral strand synthesis.
The sequence includes a region essential for binding the gene A
protein and another which is recognized for cleavage. A
consensus sequence (see Table S13-1) is shared by a number of
origins and promoters of phage DNAs. For replicative activity,
the residues marked with an asterisk are substitutable but those
marked with a circle are not. (Adapted from Dr. F. Heidekamp.)

part of a supercoiled plasmid, is not cleaved by gene A protein.[12]
These results, along with the effects of alterations in certain of the
bases of the 30-bp region, support a model in which a region removed
from the recognition sequence and cleavage site and centered between residues 4313 and 4328 is essential for an initial binding of
gene A protein to duplex DNA. This pattern is reminiscent of the
two promoter sites for RNA polymerase (Section S7-4).

Included in the essential region is a CACTAT sequence, also
encountered in the phage λ int protein binding and ori regions and
in one of the promoter-recognition sequences of fd DNA (Table
S13-1). Lack of this essential neighboring sequence in supercoiled
polyoma DNA may explain the failure of gene A protein to cleave it.
Several differences in this region may also account for resistance to
cleavage of the L-strand origin of replication of mouse mitochondrial
DNA[13] despite a remarkable 60 percent homology to the 30-bp sequence; substitution at an essential position in the cleavage site
(C for T at residue 4307, Fig. S13-1) is a further reason.

(iii) _Duplex phages_. The sequence of the principal T7 origin is
known but how it is used by T7 RNA polymerase is not (Section
S14-5). Phage T4 may have as many as four origins, and their recognition by a T4 topoisomerase is thought to open the region for the
start of DNA synthesis (Section S14-6). The sequenced origin of

12. Heidekamp, F., Baas, P. D., van Boom, J. H., Veeneman, G. H., Zipursky, S. L., and Jansz, H. S.
(1981) _N. A. Res. 9,_ 3335.
13. Doda, J. N., and Clayton, D. A. (1981) _Plasmid 6,_ 354.

TABLE S13-1

A consensus sequence in origins and promoters of phage DNAs[a]

| Phage site | Sequence[b] | Protein bound |
|---|---|---|
| $\phi$X-ori | CAACTTG↓ATATTAATAACACTATAGACCAC | gene A protein |
| λ-ori | caAAagACACTATtacaaAa | gene O protein |
| M13 (fd, f1)-ori | aagAgtcCACTATt↓aAagAa | gene 2 protein |
| λ-*att* site | aTgcagtCACTATgaAtCAa | int protein |
| M13 (fd, f1)-promoter | aTAAattCACTATtGACtct | RNA polymerase |
| T3/T7-promoter | TacgactCACTATAGggaga | RNA polymerase |
| CONSENSUS | CACTAT | |

[a]Adapted from Heidekamp, F., thesis, University of Utrecht, Netherlands (1981).
[b]The arrows indicate the positions of the $\phi$X gene A protein and the M13 gene 2 protein cleavage sites. Nucleotides homologous with part of the $\phi$X origin region are in uppercase letters; nonhomologous are in lowercase.

phage λ (Fig. 13-10, p. 462) is bound by O protein, but no more is known about its function as yet (Section S14-7).

*Plasmid origins.* How the known sequences of several plasmid origins are used for the start of synthesis and its regulation is considered in Section S14-9. For the best understood case of ColE1 plasmid, an RNA transcript of a region about 400 bp removed from the origin of the leading strand controls the frequency of initiations. Near this origin, two protein n' recognition and primosome assembly sites have also been identified (Section S11-9). One site on the L strand, downstream from the origin, may be the start of H strand synthesis for conjugal transfer; the other site on the H strand upstream from the origin may be the start of lagging strand synthesis.

*E. coli origin (oriC).* Extensive genetic manipulations of *oriC* transferred to phage and plasmid vectors have sharpened our appreciation of its organization and functions.[14,15] A unique sequence of 245 bp in the *E. coli* genome ($4 \times 10^6$ bp) is the site from which bidirectional replication is initiated (Fig. S13-2). *OriC* can be deleted from the *E. coli* chromosome and replaced either by an F or R plasmid origin or by an *oriC* sequence that supports the autonomous replication of a plasmid to which it had been transferred originally.

The *oriC* sequence is strongly conserved in *S. typhimurium* and a

14. Hirota, Y., Yamada, M., Nishimura, A., Oka, A., Sugimoto, K., Asada, K., and Takanami, M. (1981) *Prog. N. A. Res. 26*, 33.
15. von Meyenburg, K., and Hansen, F. G. (1980) *ICN–UCLA Symposia 19*, 137.

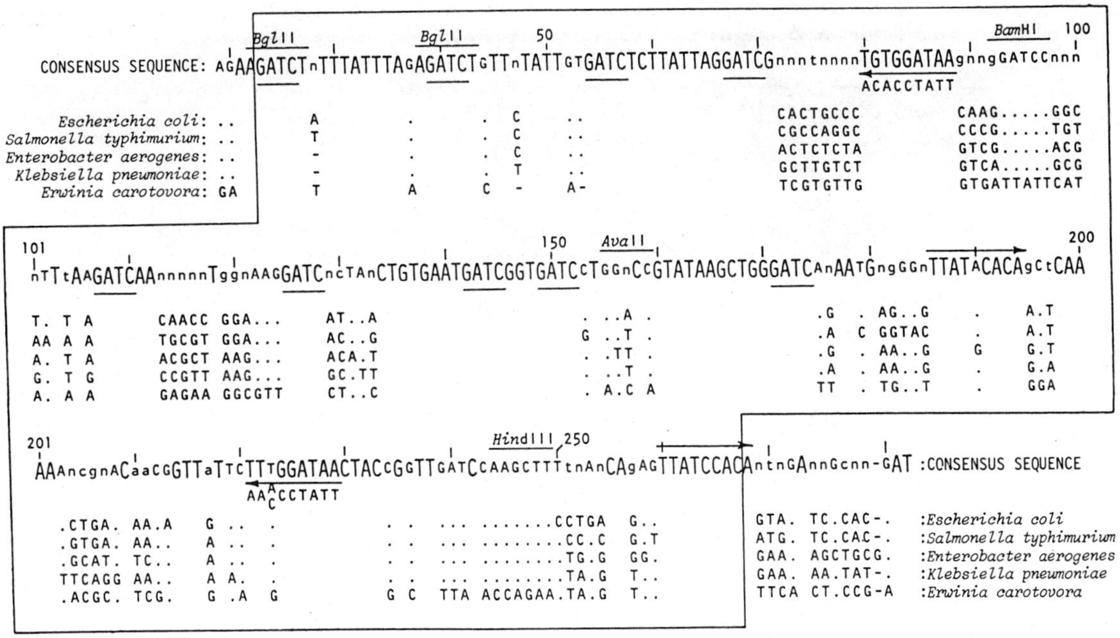

```
                                  BglII                    BglII              50                                           BamHI    100
                                  |                        |                  |              |             |               |        |
CONSENSUS SEQUENCE: AGAAGATCTnTTTATTTAGAGATCTGTTnTATTGTGATCTCTTATTAGGATCGnnntnnnnTGTGGATAAgnnggGATCCnnn
                                                                                                          ACACCTATT

    Escherichia coli: ..          A          .       . C       ..          CACTGCCC          CAAG.....GGC
Salmonella typhimurium: ..        T          .       . C       ..          CGCCAGGC          CCCG.....TGT
 Enterobacter aerogenes: ..       -          .       . C       ..          ACTCTCTA          GTCG.....ACG
   Klebsiella pneumoniae: ..      -          .       . T       ..          GCTTGTCT          GTCA.....GCG
    Erwinia carotovora: GA        T          A       C   -     A-          TCGTGTTG          GTGATTATTCAT
```

```
101                               |                150  AvaII  |               |                           200
|                                 |                   |        |               |                           |
nTTtAAGATCAAnnnnnTggnAAGGATCncTAnCTGTGAATGATCGGTGATCcTGGnCcGTATAAGCTGGGATCAnAATGnggGnTTATACACAgCtCAA

T. T A        CAACC GGA...      AT..A            . ..A .          .G  . AG..G        .  A.T
AA A A        TGCGT GGA...      AC..G       G ..T .                .A  C GGTAC       .  A.T
A. T A        ACGCT AAG...      ACA.T            . .TT .           .G  . AA..G      G  G.T
G. T G        CCGTT AAG...      GC.TT            . ..T .           .A  . AA..G      .  G.A
A. A A        GAGAA GGCGTT      CT..C       . A.C A                TT  . TG..T       .  GGA
```

```
201                               HindIII  250
|                                 |        |
AAAncgnACaacGGTTaTtcTTTGGATAACTACCGgTTGATCCAAGCTTTtnAnCAgAGTTATCCACAntnGAnnGcnn-GAT :CONSENSUS SEQUENCE
                    AAcCCTATT

.CTGA. AA.A    G ..  .          .  .  ... .......CCTGA  G..        GTA. TC.CAC-.    :Escherichia coli
.GTGA. AA..    A ..  .          .  .  ... .......CC.C   G.T        ATG. TC.CAC-.    :Salmonella typhimurium
.GCAT. TC..    A ..  .          .  .  ... .......TG.G   GG.        GAA. AGCTGCG.    :Enterobacter aerogenes
TTCAGG AA..    A A.  .          .  .  ... .......TA.G   T..        GAA. AA.TAT-.    :Klebsiella pneumoniae
.ACGC. TCG.    G .A  G          G C  TTA ACCAGAA.TA.G   T..        TTCA CT.CCG-A    :Erwinia carotovora
```

FIGURE S13-2

Consensus sequence of the minimal origin of enteric bacterial chromosomes. The consensus sequence is derived from five bacterial origin sequences, E. coli, Salmonella typhimurium, Enterobacter aerogenes, Klebsiella pneumoniae, and Erwinia carotovora. In the consensus sequence, a large capital letter means that the same nucleotide is found in all five origins; a small capital letter means the nucleotide is present in four of the five sequences; a lowercase letter is used when that nucleotide is present in three of five bacterial origins but only two different nucleotides are found at that site; and where three or four of the four possible nucleotides, or two different nucleotides plus a deletion, are found at a site, the letter n is used. In the individual origin sequences, − means a deletion relative to the consensus sequence is present, and • indicates that the nucleotide in the bacterial sequence is the same as the nucleotide in the consensus sequence. GATC sites are underlined in the consensus sequence and representative restriction sites are noted. The minimal origin of E. coli is enclosed within the box. The numbering of the nucleotide positions is that first used for E. coli, and the upper left end is the 5′ end. The four related 9-bp repeats are indicated by the arrows, with the 5′ to 3′ sequence given below the arrows. (Courtesy of Dr. J. Zyskind.)

number of other Gram-negative species (Fig. S13-2).[16] No essential genes are found within 2.2 kb up to asn (asparagine synthetase) to the right and 8.5 kb to unc or atp (ATP-synthetase operon) to the left (Fig. S13-3).[17] Removal of the region to the right of oriC was reported

16. Zyskind, J. W., and Smith, D. W. (1980) PNAS 77, 2460; Zyskind, J. W., Harding, N. E., Takeda, Y., Cleary, J. M., and Smith, D. W. (1981) ICN–UCLA Symposia 22, 13; Zyskind, J. W., Smith, D. W., Hirota, Y., and Takanami, M. (1981) ICN–UCLA Symposia 22, 26.
17. von Meyenburg, K., and Hansen, F. G. (1980) ICN–UCLA Symposia 19, 137.

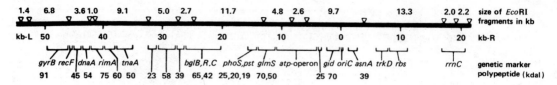

Figure S13-3
Genetic and physical map of the *oriC* region of the *E. coli* K-12 chromosome. The scale is marked with zero at the left limit of *oriC*. For details, see reference 15. (Courtesy of Professor K. von Meyenburg.)

to cause uni- rather than bidirectional replication in vivo,[18] but a comparable deletion did not interfere with bidirectional replication in vitro.[19] Coding regions exist for a 21-kdal protein just to the left of *oriC* and for a 16-kdal protein to the right. Grossly, the *oriC* site of 245 bp contains the essential genetic information for replicator function and sequences to either side do not have a profound effect either on replication in vivo or on cell division. Nevertheless, features that are difficult to measure with precision, such as copy number, may be significantly influenced by neighboring regions.[20]

*OriC* function is readily assessed by autonomous replication of a plasmid (e.g., ColE1) conditionally deprived of its own origin function in a mutant host. Plasmid multiplication is assayed by a gene function (e.g., antibiotic resistance) in a *polA*+ as compared to a *polA*- host. Deletions, insertions, and substitutions of one or more base pairs within *oriC* show (Fig. S13-4):[21] (i) Certain bases are essential and others are not, indicating the need to recognize several particular sequences, and (ii) regions of defined length rather than sequence are needed as spacers to fix a distance between regions with essential sequences.

Several other features are notable in the *oriC* sequence. Two back-to-back promoters may be loci for RNA polymerase in DNA initiation rather than being connected with mRNA synthesis.[22] The GATC sequence that appears fourteen times in the 300-bp *oriC* region (Fig. S13-2, S13-4) instead of the statistical frequency of about once is a *dam* methylase site, but the normal growth of *dam* mutants is evidence against an essential methyl acceptor role for the GATC sequences. The lack of symmetrical sequences for a bidirectional start and the location of two RNA–DNA junctions of nascent frag-

18. Meijer, M., and Messer, W. J. (1980) *J. Bact. 143*, 1049.
19. Kaguni, J. M., Fuller, R., and Kornberg, A. (1982) *Nat. 296*, 623.
20. Ogura, T., Miki, T., and Hiraga, S. (1980) *PNAS 77*, 3993.
21. Hirota, Y., Oka, A., Sugimoto, K., Asada, K., Sasaki, H., and Takanami, M. (1981) *ICN–UCLA Symposia 22*, 1.
22. Lother, H., and Messer, W. (1981) *Nat. 294*, 376.

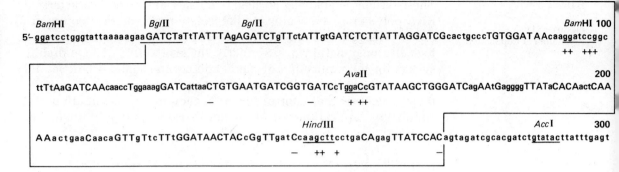

FIGURE S13-4

Base-substitution mutations within *oriC*. Capital letters represent conservative (consensus) residues, lowercase letters nonconservative residues deduced by comparison of sequences of five bacterial origins. (+) indicates mutations that show no effect, and (−) indicates mutations that result in defective *ori* function. (Courtesy of Dr. M. Takanami and Dr. Y. Hirota.)

ments on the same *oriC* strand[23] suggest that initiation is an asymmetric process. Several potential secondary structures exist in the *oriC* sequence,[24] but enthusiasm for their significance is diminished by equally attractive branched and stem-and-loop structures outside the *oriC* region.

The numerous analyses of *oriC* emphasize that recognition and spacer sequences must be strictly maintained. Whether these rules apply in vitro can be examined with the oriC enzyme system (Section S11-15). Degrees of replication insufficient to maintain the plasmid in vivo may be detected by the enzyme system. Several proteins should be needed to recognize and to exploit the sequences and topological features of the 245-bp *oriC* site. Their organization into an elaborate entity, with RNA polymerase as only one of the many components, can be anticipated. Although the minimal origin sequence suffices to allow autonomous plasmid replication, adjacent sequences may contribute some subtle functions of the more complex behavior expected of the chromosome.

Ultimately, the exploitation of the oriC system in vitro (Section S11-15) should allow a direct biochemical approach to the regulation of replication. By way of example, the present understanding of the regulation of gene expression in bacteria has depended, to a large extent, upon the establishment of in vitro transcription systems.

23. Okazaki, T., Hirose, S., Ogawa, T., Fujiyama, A., and Kohara, Y. (1980) *ICN–UCLA Symposia* **19**, 429.
24. Hirota, Y., Yamada, M., Nishimura, A., Oka, A., Sugimoto, K., Asada, K., and Takanami, M. (1981) *Prog. N. A. Res.* **26**, 33.

These systems, employing purified templates and a well-characterized RNA polymerase, have allowed the isolation of factors affecting initiation, elongation, and termination of specific transcripts in physiologically meaningful ways. Similarly, the availability of an authentic system for the replication of the E. coli chromosomal origin should allow the isolation of those factors, as well as the understanding of those processes that control the frequency of origin-initiation and couple the duplication rate of the chromosome to that of the cell as a whole.

*B. subtilis origin.*[25]  Unlike *oriC,* the cloned *B. subtilis* origin contains a suppressor sequence, transcription from which inhibits cellular growth and replication of the plasmid itself. A 5.7-kb region (B7) that includes the origin of the *B. subtilis* chromosome confers unusual properties on a plasmid that normally replicates autonomously both in *E. coli* and *B. subtilis.* A 2.2-kb section of B7 is responsible for inhibitory effects on the host cell, presumably by encoding a protein that in *E. coli* causes the cell surface to become sticky and in *B. subtilis* depresses growth, sporulation, and viability. In a terminal portion of the 2.2-kb section is a 490-bp sequence that has a profound *cis*-inhibitory effect on replication of the plasmid itself. These properties suggest a mechanism of autorepression of initiation during a cycle of replication in *B. subtilis* that must be relieved by a change in secondary or tertiary structure in or near the 490-bp sequence. As judged by hybridization, the 2.2-kb section is present in at least seven different sites on the *B. subtilis* chromosome indicating further intricacies in the control of initiation and the termination of replication in *B. subtilis.* Isolation of RNA that appears to be covalently linked to chromosomal DNA[26] and is synthesized at the onset of *B. subtilis* replication cycles suggests still another complexity in the initiation process.

## S13-6  Eukaryotic Replication: Control

The pace of replication is regulated at three levels: the cell cycle, replicons of about 30 kb within a chromosome, and fork movement. At none of these are molecular mechanisms defined. Experimental problems are more complicated than with prokaryotes because of the profound effects of differentiation on a cell line and on cells within a multicellular organism.[27] There is also the concern that as models, animal viruses emancipated from cellular restraints may not

---

25. Seiki, M., Ogasawara, N., and Yoshikawa, H. (1981) *MGG 183,* 220; 227.
26. Henckes, G., Vannier, F., Buu, A., and Seror-Laurent, S. J. (1982) *J. Bact. 149,* 79.
27. Levenson, R., and Housman, D. (1981) *Cell 25,* 5.

be as illuminating as phages have been.[28] The most promising leads may be provided by studies of replication of autonomous plasmids in yeast (also Section S15-8) and of DNA injected into the *Xenopus* egg.

## The Cell Cycle

The life cycle of the cell depends on interactions between the *chromosome cycle* in which chromosomes are replicated and segregated and the *growth cycle* in which cell size and components are doubled. How these two cycles are coordinated is still a mystery. The chromosome cycle is set in motion in a preparatory interval ($G_1$); it then spans periods of DNA synthesis (S), preparation for mitosis ($G_2$), and mitosis (M). During the lengthy $G_1$ phase that bridges the gap between states of nuclear quiescence ($G_0$) and mitogenesis, products are presumably generated that commit the cell to proliferation. However, speculations about the nature of these processes[29] must contend with the fact that most of the $G_1$ phase is essentially part of the growth cycle and can be eliminated from the chromosome cycle simply by lengthening the S phase.[30]

## Origins

Chromosomal origins in eukaryotes differ from those of prokaryotic genomes in several ways.

(i)  Origins are multiple, about one per 20 kb strung along each of the multiple chromosomes (Fig. 11-11, p. 360), rather than being limited to one or a few in a single chromosome.

(ii)  Origins are not reinitiated before completion of the chromosome cycle as they are in dichotomous replication in a rapidly growing prokaryote. However, some substances might cause illegitimate reinitiations ("replicon misfirings") with major metabolic and evolutionary consequences.[31] Phorbol esters (e.g., 12-O-tetradecanoylphorbol 13-acetate) potent in promoting tumors and gene amplification[32] may prove to be an example of such reinitiating activity at an origin locus.

(iii)  Origins need not be fixed in sequence or location. Foreign DNA (circular SV40 or bacterial plasmids) is replicated when injected into frog eggs;[33] replication is synchronous with the chromo-

---

28. De Pamphilis, M. L., and Wassarman, P. (1980) *ARB 49,* 627.
29. Das, M. (1981) *PNAS 78,* 5677.
30. Stancel, G. M., Prescott, D. M., and Liskay, R. M. (1981) *PNAS 78,* 6295.
31. Varshavsky, A. (1981) *PNAS 78,* 3673.
32. Varshavsky, A. (1981) *Cell 25,* 561.
33. Laskey, R. A., and Harland, R. M. (1981) *Cell 24,* 283.

some cycle and presumably responds to the same regulatory controls. Nevertheless, strong indications for defined origins in eukaryotes do exist.

Certain sequences called ars (autonomous replicating sequence), most of which differ in composition, are present in about 400 places in the yeast genome, a frequency consistent with the number of origins observed microscopically. Ars sequences confer on plasmids the capacity for autonomous replication in yeast; centromeric sequences (cen) provide plasmids with a more stable extrachromosomal maintenance through mitotic and meiotic cycles.[34] Ars sequences are widely distributed among eukaryotic genomes, including mammalian DNAs, and significantly, appear to be absent from E. coli DNA.[34,35] Plasmids that include yeast ars sequences require the functions of cdc (cell-development-cycle) yeast genes, and the replication of these plasmids in yeast cells and extracts is initiated from a defined origin (Section S15-8). The fixed origins for unidirectional replication in mitochondria are well established (Section S15-8), and such Xenopus mitochondrial regions cloned in plasmids support high frequency, stable transformation in yeast.[36] This property is not displayed by certain other regions of Xenopus DNA, or repetitive human DNA or SV40 DNA.

The permissiveness of frog eggs in replicating foreign DNA without ars sequences may prove atypical and quantitatively comparable to the use of secondary origins in phages and to the substitution of phage and plasmid origins for the unique E. coli origin. The distribution and properties of ars sequences are clearly consistent with the features expected of replication origins and can be analyzed further with available genetic and biochemical systems (Section S15-8). The promotional signals for eukaryotic replication may eventually resemble in variety the increasingly complex assortment of signals and factors regulating the starts of eukaryotic transcription.

Transformation to Proliferation[37]

The products of viral oncogenes responsible for initiation and maintenance of malignant transformation are also formed by the uninfected quiescent cell. The transforming src gene of Rous sarcoma virus encodes a 60-kdal kinase that specifically phosphorylates

34. Stinchcomb, D. T., Thomas, M., Kelly, J., Selker, E., and Davis, R. W. (1980) PNAS 77, 4559; Chan, C. S. M., and Tye, B.-K. (1980) PNAS 77, 6329; Beach, D., Piper, M., and Shall, S. (1980) Nat. 284, 185; Hsiao, C.-L., and Carbon, J. (1981) PNAS 78, 3760.
35. Kiss, G. B., and Pearlman, R. E. (1981) Gene 13, 281.
36. Zakian, V. A. (1981) PNAS 78, 3128.
37. Klein, G. (1981) Nat. 294, 313; Bishop, J. M. (1981) Cell 23, 5.

tyrosines of certain proteins, such as a 34- to 36-kdal protein and vinculin. Normal vertebrate cells contain loci related to every retrovirus oncogene known and encode highly similar protein kinases. The essence of transformation may be the action of a more powerful promoter for transcription of the cellular *sarc* gene (cellular oncogene or protooncogene) and the consequent overproduction of a key regulatory enzyme. How the pleiotropic changes of the transformed cell result from alterations in the level of the viral (*src*) or cellular (*sarc*) gene products is a key question.

## Selective Gene Amplification[38]

In at least two cases, gene amplification is used in terminally differentiated cells to achieve high level expression of mRNAs and proteins. In *Drosophila melanogaster,* genes responsible for continuous high level synthesis of the chorion proteins are amplified about 50-fold in follicle cells by selective hyper-replication, resulting in a partially polytene structure extending over about 100 kb.[39] In developing chick myotubules, the $\alpha$-actin genes (and probably other muscle-specific genes as well) are amplified about 100-fold, but only during the 2-day period of rapid muscle protein synthesis. Here, rapid formation and loss of extrachromosomal templates for transcription seem likely. These templates may be circles or linear tandem repeats produced by rolling-circle replication.[40] Novel DNA joints found in the amplified DNA support such a mechanism, which is similar to that used for high level synthesis of rRNA in amphibian oocytes.[41]

*Gene amplification revealed by drug resistance.* In some cell lines resistant to methotrexate, double-minute chromosomes carry the amplified dihydrofolate reductase genes.[42] Although the detailed structure of double-minute chromosomes is unknown, it seems likely that they are circular or linear tandem repeats of cellular sequences. Such chromosomal units are also found consistently in several different types of tumors, especially neuroblastomas and retinoblastomas, inviting the speculation that amplification and consequent overproduction of one or more proteins may be related to tumorigenesis in such cases.

---

38. *Gene Amplification* (Schimke, R. T., ed.) (1982) CSHL.
39. Spradling, A. C. (1981) *Cell 27,* 193.
40. Zimmer, W. E., Jr., and Schwartz, R. J. (1982) in *Gene Amplification,* CSHL, p. 137.
41. Brown, D. D., and David, I. B. (1968) *Science 160,* 272.
42. Schimke, R. T., Brown, P. C., Kaufman, R. J., McGrogan, M., and Slate, D. L. (1980) *CSHS 45,* 785; Haber, D. A., and Schimke, R. T. (1981) *Cell 26,* 355.

In other examples of resistance to methotrexate,[43] N-phosphonacetyl L-aspartate,[44] and cadmium ions,[45] the respective amplified genes for the overproduced dihydrofolate reductase, CAD (first three enzymes of pyrimidine biosynthesis), and metallothionein I are arranged linearly over a considerable length of chromosomal DNA, with 100–500 kb of amplified DNA present per gene. In one case these sequences replicate synchronously, early in S phase.[46] Much of the amplified DNA surrounding the CAD gene has been cloned and novel joints are present. The DNA sequences of such joints, together with information on how often each portion of the DNA is amplified in a set of independent events, should give clues about the mechanism of amplification. Simple duplication by recombinational events between short regions of homology seems very unlikely, and the data for the chromosomal amplifications are not consistent with onion-skin (polytenic) structures. However, formation of these structures, followed by recombinational events to introduce the repeated sequences into the linear chromosome, is possible, as is introduction of tandemly repeated sequences into the chromosome by transposons.[47]

43. Biedler, J. L., and Spengler, B. A. (1976) *Science 191*, 185; Milbrandt, J. D., Heintz, N. H., White, W. C., Rothman, S. M., and Hamlin, J. L. (1981) *PNAS 78*, 6043.
44. Wahl, G. M., Vitto, L., Padgett, R. A., and Stark, G. R. (1982) *Mol. Cell. Biol. 2*, in press.
45. Mayo, K. E., and Palmiter, R. D. (1982) in *Gene Amplification* (Schimke, R. T., ed.) CSHL, p. 67; Beach, L. R., and Palmiter, R. D. (1981) *PNAS 78*, 2110.
46. Milbrandt, J. D., Heintz, N. H., White, W. C., Rothman, S. M., and Hamlin, J. L. (1981) *PNAS 78*, 6043.
47. Harshey, R. M., and Bukhari, A. I. (1981) *PNAS 78*, 1090.

# S14

# Bacterial DNA Viruses and Plasmids

## Abstract

Despite severe financial malnutrition, phage studies continue to produce fundamental insights into replication mechanisms and the cellular replication process. Were it not for the incidental utility of phages as vectors for recombinant DNA, for the relationships of phages to the popular plasmids and transposable elements, and for the obvious utility of animal viruses in eukaryotic cellular biology, phage research would be nearly starved by now.

New information about the structure of filamentous phages makes the connections between phage adsorption and penetration more mysterious. The role for an adsorption protein to "pilot" (p. 476) the phage DNA through membranes to a cellular berth for DNA expression and replication remains an attractive notion with no new facts to support or deny it.

Of all replicative cycles in nature, replication of the phage genomes are the best known in molecular detail. Conservation of the host primosome by $\phi$X174, the powerful helicase-primase action of gene 4 protein of T7, and the elaborate helicase-primase-polymerase system of T4 represent interesting variations on a basic theme of replication fork movement along a chromosome. Understanding of phage $\lambda$ replication has lagged behind that of the virulent phages but promises to catch up now that the elusive O and P proteins have been cloned and overproduced.

Plasmids occupy center stage as models for examining the initiation of a cycle of chromosome replication and how it is controlled. A small RNA whose level can regulate starts of DNA synthesis at the origin of replication is a device both novel and familiar. Using RNA rather than protein for specific recognition of a DNA site was not anticipated; on the other hand, RNA hybridization and secondary structure are devices used in the control of transcription by attenuation. Discovery of several in vitro systems for plasmid DNA replication should soon make the molecular details of these processes as clearly known as those of the phages.

## S14-3   Small Filamentous Phages: M13, fd, f1

Virion

In addition to the 2700 copies of the gene 8 coat protein (B) that ensheathes the long loop of single-stranded DNA, the virion contains about five copies each of the products of genes 3, 6, 7, and 9 (Fig. S14-1).[1] Gene 3 and 6 proteins (A and D) make up the adsorption complex which appears as a knob and stem complex at one end of the phage while the gene 7 and 9 proteins (C) are at the other end.[2] Surprisingly, the DNA is oriented with the region containing the complementary strand origin at the end of the virion opposite the adsorption complex.[3]

_Coat._   The major coat proteins of all the filamentous phages (fd, f1, M13, ZJ-2, $1f_1$, Ike) have a similar structure: a central hydrophobic region flanked by positively and negatively charged residues (p. 481).[4]

1. Simons, G. F. M., Konings, R. N. H., and Schoenmakers, J. G. G. (1979) _FEBS Lett. 106_, 8; (1981) _PNAS 78_, 4194; Lin, T.-C., Webster, R. E., and Konigsberg, W. (1980) _JBC 255_, 10331.
2. Grant, R. A., Lin, T.-C., Konigsberg, W., and Webster, R. E. (1981) _JBC 256_, 539; Gray, G. W., Brown, R. S., and Marvin, D. A. (1981) _JMB 146_, 621.
3. Webster, R. E., Grant, R. A., and Hamilton, L. A. W. (1981) _JMB 152_, 357.
4. Nakashima, Y., Frangione, B., Wiseman, R. L., and Konigsberg, W. H. (1981) _JBC 256_, 5792.

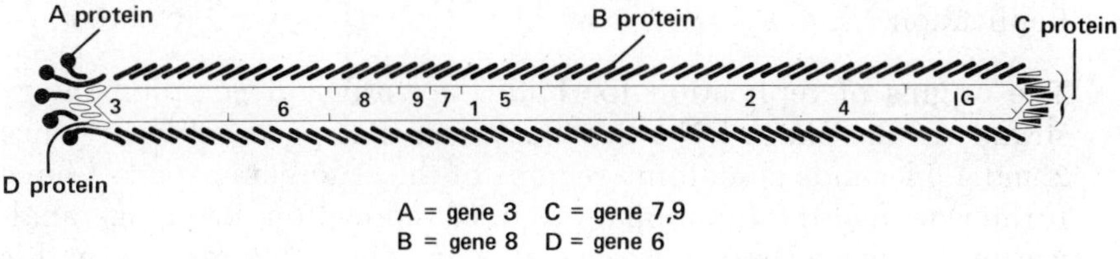

A = gene 3  C = gene 7,9
B = gene 8  D = gene 6

FIGURE S14-1
Schematic representation of the coat proteins and genetic map of a filamentous phage
(e.g., M13, fd). Designation of genes by roman numerals is also in common usage. IG
denotes the intergenic sequence that contains the origin for starting synthesis of the
complementary strand. (Courtesy of Professor R. E. Webster.)

In the infected cell, the gene 8 protein spans the cytoplasmic membrane, with the amino-terminal portion exposed on the outside and
the carboxy-terminal amino acids located on the cytoplasmic surface.[5] Involvement of a specific endopeptidase, a leader peptidase
that removes the amino-terminal signal polypeptide in vitro and in
vivo,[6] may serve as a model for biosynthesis of membrane proteins
in prokaryotes and eukaryotes.

Adsorption and Penetration

Intact gene 3 protein is essential for adsorption to F-piliated *E. coli*.[7]
Proteolytic removal of the knob portion of the protein renders the
phage noninfective.[8] The production of some phage proteins, such as
gene 2 protein, confers partial resistance to further infection by the
phage.[9] Were the penetration of phage DNA into the cell tightly
coupled to synthesis of the complementary strand (p. 487), one would
expect the replication origin on the viral strand to take the lead. Yet
this DNA region appears to be remote from the adsorption complex
(Fig. S14-1) in normal or altered phage particles.[10] Perhaps, as a result
of the kind of structural changes in the virion produced by organic
solvents, contact with the bacterial membrane causes the major
coat protein to dissolve in the membrane and extrude the DNA in
an oriented manner into the cytoplasm.[11]

5. Ohkawa, I., and Webster, R. E. (1981) *JBC 256*, 9951.
6. Zwizinski, C., Date, T., and Wickner, W. (1981) *JBC 256*, 3593; Date, T., and Wickner, W. (1981)
   *PNAS 78*, 6106; Boeke, J. D., Russel, M., and Model, P. (1980) *JMB 144*, 103; Russel, M., and
   Model, P. (1981) *PNAS 78*, 1717.
7. Nelson, F. K., Friedman, S. M., and Smith, G. P. (1981) *Virol. 108*, 338.
8. Grant, R. A., Lin, T.-C., Konigsberg, W., and Webster, R. E. (1981) *JBC 256*, 539; Gray, G. W.,
   Brown, R. S., and Marvin, D. A. (1981) *JMB 146*, 621.
9. Dotto, G. P., Enea, V., and Zinder, N. D. (1981) *PNAS 78*, 5421.
10. Webster, R. E., Grant, R. A., and Hamilton, L. A. W. (1981) *JMB 152*, 357; Nakashima, Y.,
    Frangione, B., Wiseman, R. L., and Konigsberg, W. H. (1981) *JBC 256*, 5792.
11. Griffith, J., Manning, M., and Dunn, K. (1981) *Cell 23*, 747; Manning, M., Chrysogelos, S., and
    Griffith, J. (1981) *J. Virol. 40*, 912.

The origins of replication for both the viral and complementary strands of the phage DNA are located in the space between genes 2 and 4. Plasmids containing regions of this intergenic space[12] have further defined the functional origin of replication. Sequence analysis of the phage DNA suggests stable hairpin structures within this origin region for complementary strand synthesis (pp. 479, 482), one of which causes phage T4 polymerase to pause when it replicates the viral DNA.[13]

RF replication proceeds effectively in soluble extracts of infected cells supplemented with single-strand binding protein.[14] Synthesis with a reconstituted enzyme system is initiated by the introduction of a nick at a specific site on the viral strand (viral strand origin) of the RF molecule by the gene 2 protein (p. 492). As in the rolling-circle mechanisms of φX174 (Section S11-11), replication of the viral strand proceeds until it reaches the regenerated viral strand origin where, it may be inferred from studies in vivo,[15] gene 2 protein nicks again and generates a unit-length, circular viral strand. Gene 2 cloned in a high-copy-number plasmid is overproduced 200-fold and acts in *trans* on other plasmids engineered to depend on the fd origin for their replication.[16]

As the concentration of gene 5 protein increases, a transition from RF synthesis to single-strand accumulation occurs and gene 5 protein-DNA complexes become abundant. The complex appears to be a spindle or core of gene 5 protein molecules around which the DNA molecule is spooled.[17] Gene 5 protein effects a switch from RF to viral DNA synthesis by inhibiting synthesis of gene 2 protein (see below) as well as by diverting the viral DNA from its template role.

Gene Expression

Ten proteins are encoded by the phage genome. The gene X product is in phase with and overlaps gene 2,[18] as previously predicted from DNA sequence and in vitro transcription-translation. In vivo, the

12. Cleary, J. M., and Ray, D. S. (1980) *PNAS 77*, 4638; Dotto, G. P., Enea, V., and Zinder, N. D. (1981) *Virol. 114*, 463.
13. Huang, C.-C., and Hearst, J. E. (1980) *Anal. Biochem. 103*, 127; Huang, C.-C., Hearst, J. E., and Alberts, B. M. (1981) *JBC 256*, 4087.
14. Van Dorp, B., Schneck, P. K., and Staudenbauer, W. L. (1979) *EJB 94*, 445.
15. Horiuchi, K. (1980) *PNAS 77*, 5226; Meyer, T. F., Bäumel, I., Geider, K., and Bedinger, P. (1981) *JBC 256*, 5810; Harth, G., Bäumel, J., Meyer, T. F., and Geider, K. (1981) *EJB 119*, 663.
16. Meyer, T. F., and Geider, K. (1981) *PNAS 78*, 5416; Dotto, G. P., Enea, V., and Zinder, N. D. (1981) *PNAS 78*, 5421.
17. McPherson, A., Jurnak, F., Wang, A., Kolpak, F., Rich, A., Molineux, J., and Fitzgerald, P. (1980) *Biophys. J. 32*, 155; Torbet, J., Gray, D. M., Gray, G. W., Marvin, D. A., and Siegrist, H. (1981) *JMB 146*, 305.
18. Yen, T. S. B., and Webster, R. E. (1981) *JBC 256*, 11259.

proteins appear to be synthesized from two classes of transcripts. The messages in the most abundant class start at different strong promoters located between the intergenic space and gene 8, but all end at the single rho-independent termination site just beyond gene 8 (Fig. 14-3).[19] The longer transcripts with genes 2 and X appear to be processed rapidly to long-lived mRNAs containing genes 5 and 8, thus accounting for the high level of synthesis of these gene products.[20] The remaining transcripts are synthesized in low amounts from weak promoters located at various points on the remainder of the genome and apparently end at a rho-dependent terminator in the intergenic space. Gene 2 protein is specifically overproduced in infected bacteria containing a nonfunctional gene 5 protein,[21] suggesting an inhibition by gene 5 protein of gene 2 protein synthesis at the translational level.[22]

## Phage Assembly

A 200- to 300-nucleotide region at the junction of gene 4 and the intergenic space is necessary for efficient packaging of the DNA into particles.[23] How the morphogenetic process is set in motion and the mature virion assembled at the membrane are still unclear.

# S14-4   Small Polyhedral Phages:
$\phi$X174, S13, G4

Enzymology of replication, structure-function relationship of the origin of viral strand replication, and in vitro phage assembly have been the principal subjects of recent studies.

## Genome

Of eleven proteins encoded by the compact $\phi$X174 genome (Fig. S14-2), only gene K protein appears to be nonessential, although mutants do show some reduction in burst size.[24] A proposed role for the abbreviated gene A protein (gene A*) in the shutoff of host DNA syn-

19. Cashman, J. S., Webster, R. E., and Steege, D. A. (1980) *JBC 255*, 2554; Smits, M. A., Schoenmakers, J. G. G., and Konings, R. N. H. (1980) *EJB 112*, 309; LaFarina, M., and Model, P., personal communication.
20. Cashman, J. S., Webster, R. E., and Steege, D. A. (1980) *JBC 255*, 2554; Smits, M. A., Schoenmakers, J. G. G., and Konings, R. N. H. (1980) *EJB 112*, 309; LaFarina, M., and Model, P., personal communication.
21. Meyer, T. F., and Geider, K. (1979) *JBC 254*, 12636; Webster, R. E., and Rementer, M. (1980) *JMB 139*, 393.
22. Model, P., McGill, C., and Mazur, B. J., personal communication; Yen, T. S. B., and Webster, R. E., personal communication.
23. Webster, R. E., Grant, R. A., and Hamilton, L. A. W. (1981) *JMB 152*, 357; Dotto, G. P., Enea, V., and Zinder, N. D. (1981) *Virol. 114*, 463.
24. Tessman, E. S., Tessman, I., and Pollock, T. J. (1980) *J. Virol. 33*, 557.

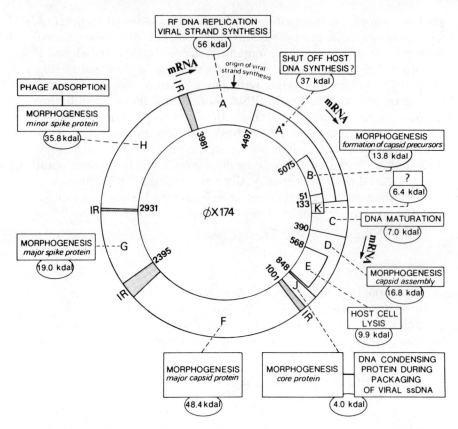

FIGURE S14-2
Genetic map of phage φX174 with suggested functions of gene products. IR, intergenic region. (Courtesy of Dr. F. Heidekamp.)

thesis (p. 505) is yet to be established. The origin sequence for viral strand synthesis (Table S13-1) is strongly conserved in this family of icosahedral phages that differ considerably in size and sequence of their genomes. Recombinant DNA plasmids possessing this origin[25] are replicated in vitro by the φX rolling-circle mechanism (Section 11-11) and can provide a superior source of single-stranded, radioactive DNA as probes for hybridization measurements.

Replication

Of all replicative cycles in nature, replication of the genomes of this phage family is best known in molecular detail. The three major stages of the replicative cycle (p. 506) have been largely reconstituted with pure proteins.

25. Heidekamp, F., Baas, P. D., van Boom, J. H., Veeneman, G. H., Zipursky, S. L., and Jansz, H. S. (1981) N. A. Res. 9, 3335.

*SS→RF.* Conversion of the infecting viral strand to the replicative form by host proteins organized as a primosome serves as a model for the priming events in the discontinuous phase of host chromosome replication (Sections S11-9, S11-10). The primosome, either assembled on the viral strand or appropriated from the host replication fork, remains complexed with the completed RF.

*RF→SS→RF.* In RF multiplication, viral strands are synthesized continuously by the rolling-circle mechanism (Section S11-11) initiated by cleavage by gene A protein (the only phage-encoded replication protein). The synthetic RF, complexed with the primosome, is split twenty times faster than the deproteinized RF isolated from infected cells, notwithstanding the low superhelical density of the in vitro product. Location of the primosome near the viral strand origin may facilitate cleavage by the gene A protein. Processive primosome movement at the replication fork as with the host chromosome (Section S11-14) primes the synthesis of a complementary strand to convert the viral strand to RF. (It is still uncertain whether complementary strand synthesis is initiated before the unit-length viral strand is completed.) All these findings suggest that the parental RF with its *conserved primosome* is the only RF replicating unit and that the function of the numerous supercoiled RFs in the infected cell is transcriptional.

The functions and similarity of the pure gene A proteins encoded by $\phi$X174[26] and G4[27] are well established. However, a role for the gene A* proteins in shutting off host DNA replication has not been proven by their in vitro actions.[28] The gene A* protein cleaves single-stranded $\phi$X174 viral DNA at many sites with a strong preference for the origin of replication but is inactive on the supercoiled RF substrate of the gene A protein.

*RF→SS→virions.* Encapsidation of viral strands from the RF→SS rolling circle supervenes over their conversion to RF when the required maturation and coat proteins become available. Morphogenesis of $\phi$X174 virions in vitro[29] follows a pathway similar to that inferred from in vivo studies. Purified viral components,[30] together with a crude extract from uninfected *E. coli* (to supply replication

26. Eisenberg, S., and Kornberg, A. (1979) *JBC 254,* 5328; Langeveld, S. A., van Arkel, G. A., and Weisbeek, P. J. (1980) *FEBS Lett. 114,* 269; Dubeau, L., Hours, C., and Denhardt, D. T. (1981) *Can. J. Biochem. 59,* 106.
27. Weisbeek, P., van Mansfeld, F., Kuhlemeier, C., van Arkel, G., and Langeveld, S. (1981) *EJB 114,* 501.
28. Langeveld, S. A., van Mansfeld, A. D. M., van der Ende, A., van de Pol, J. H., van Arkel, G. A., and Weisbeek, P. J. (1981) *N. A. Res. 9,* 545; Eisenberg, S., and Ascarelli, R. (1981) *N. A. Res. 9,* 1991; Dubeau, L., and Denhardt, D. T. (1981) *BBA 653,* 52.
29. Mukai, R., Hamatake, R. K., and Hayashi, M. (1979) *PNAS 76,* 4877; Koths, K., and Dressler, D. (1980) *JBC 255,* 4328.
30. Aoyama, A., Hamatake, R. K., and Hayashi, M. (1981) *PNAS 78,* 7285.

proteins, including the gyrase A subunit),[31] produce infectious phage particles. The phage-encoded proteins that are required, in addition to gene A protein, to sustain rolling-circle replication are gene F, G, H, B, and D proteins, which form the phage head precursor, and gene C and J proteins, which feed in the viral DNA and pack it properly.

## S14-5  Medium-Size Phages: T7 and Other T-Odd (T1, T3, T5)

Phage T7[32]

*Genome*.  The entire nucleotide sequence of the 40-kb genome of T7 has been determined;[33] there are 50 known or potential genes, and 35 proteins have been identified. The sequence of the genetic left end (30 percent of the genome)[34] contains, in addition to the thirteen proteins previously known to be encoded in this region,[35] sequences for eight potential proteins. Although no genes overlap within this region, the arrangement of genetic elements is efficient and includes promoters for host and T7 RNA polymerases, RNaseIII recognition sites, and the primary origin for the initiation of DNA replication.

*Infection*.  The remarkable resistance of T7 DNA to cleavage by the *E. coli* restriction system has been explained by the gene 0.3-encoded inhibitor of the host restriction enzyme. The entering DNA survives destruction during the time needed to produce the inhibitor because gene 0.3, at the head of the genome entering the cell, lacks *E. coli* restriction sites; further entry of the genome is delayed until the inhibitor is produced.

*Replication origins*.  The primary origin of phage T7 DNA replication is located within a 129-bp region between 14.75 and 15.0 percent of the distance from the left end of the genome.[36] Deletion mutants lacking the primary origin initiate replication at secondary sites, the

31. Hamatake, R. K., Mukai, R., and Hayashi, M. (1981) *PNAS 78*, 1532.
32. Kruger, D. H., and Schroeder, C. (1981) *Microbiol. Rev. 45*, 9.
33. Studier, F. W., personal communication.
34. Dunn, J. J., and Studier, F. W. (1981) *JMB 148*, 303; Stahl, S. J., and Zinn, K. (1981) *JMB 148*, 481; Tamanoi, F., Engler, M. J., Lechner, R., Orr-Weaver, T., Romano, L. J., Saito, H., Tabor, S., and Richardson, C. C. (1980) *ICN–UCLA Symposia 19*, 411; Tabor, S., Engler, M. J., Fuller, C. W., Lechner, R. L., Matson, S. W., Romano, L. J., Saito, H., Tamanoi, F., and Richardson, C. C. (1981) *ICN–UCLA Symposia 22*, 387.
35. Rosenberg, A. H., Simon, M. N., Studier, F. W., and Roberts, R. J. (1979) *JMB 135*, 907; Studier, F. W., Rosenberg, A. H., Simon, M. N., and Dunn, J. J. (1979) *JMB 135*, 917.
36. Tamanoi, F., Saito, H., and Richardson, C. C. (1980) *PNAS 77*, 2656.

end. The primary origin lies in an AT-rich intergenic region that contains a gene 4 protein (primase) recognition site, and is preceded by two tandem promoters for T7 RNA polymerase.[37]

### Replication protein

(i) _DNA polymerase_ (Section S5-6). The gene 5 protein, active only when tightly complexed with host thioredoxin, is responsible for replication of the T7 genome.

(ii) _Gene 4 protein: helicase and primase_ (Section S11-10). Acting alone, gene 4 protein is a helicase and as a primase synthesizes a tetraribonucleotide at a specific template site (e.g., CTGGG/T). Coupled with T7 polymerase and single-strand binding protein, a duplex replicating fork is generated.

(iii) _Gene 2 protein_. It is an inhibitor of host RNA polymerase and is also required for packaging DNA into phage heads,[38] perhaps for its role in the synthesis of concatameric DNA.[39] Host mutants with an RNA polymerase defective in binding gene 2 protein are deficient in DNA replication and packaging.[40]

(iv) _Single-strand binding protein_ (Section S9-5). The gene 2.5 protein, only partially replaceable by E. coli single-strand binding protein, is essential for replication.

(v) _Gene 1.2 protein_. Adjacent to the primary origin of replication are the two small genes,[41] 1.1 and 1.2. Neither gene product is required for growth of T7 in wild-type E. coli; deletion mutants of T7 lacking this region grow normally. However, an E. coli mutant (optA1) cannot support the growth of either point or deletion mutants of T7 defective in gene 1.2.[42] The optA1 mutation is at 3.6 minutes on the E. coli linkage map and shows no phenotypic effect. Mutant cells infected with the T7 gene 1.2 mutant are defective in T7 DNA replication; synthesis stops early and the newly synthesized DNA is rapidly degraded. Expression of gene 1.2 is regulated by ribonuclease III processing of mRNA.[43] The gene 1.2 protein is thus another example of a T7 phage protein (e.g., DNA ligase) that is dispensable in the presence of what is presumed to be a functionally similar host protein.

37. Saito, H., Tabor, S., Tamanoi, F., and Richardson, C. C. (1980) _PNAS 77_, 3917; Dunn, J. J., and Studier, F. W. (1981) _JMB 148_, 303.
38. LeClerc, J. E., and Richardson, C. C. (1979) _PNAS 76_, 4852.
39. De Wyngaert, M. A., and Hinkle, D. C. (1980) _J. Virol. 33_, 780.
40. De Wyngaert, M. A., and Hinkle, D. C. (1979) _JBC 254_, 11247.
41. Saito, H., Tabor, S., Tamanoi, F., and Richardson, C. C. (1980) _PNAS 77_, 3917; Dunn, J. J., and Studier, F. W. (1981) _JMB 148_, 303.
42. Saito, H., and Richardson, C. C. (1981) _J. Virol. 37_, 343.
43. Saito, H., and Richardson, C. C. (1981) _Cell 27_, 533.

*Replication with purified proteins*

(i) <u>The replication fork</u> can be partially reconstituted with two T7 phage proteins: DNA polymerase and gene 4 protein.[44] Leading strand synthesis by T7 DNA polymerase occurs only in conjunction with the gene 4 protein (Fig. S11-9). The latter binds and migrates 5′ to 3′ on the displaced strand, facilitating the unwinding of the duplex. When a recognition site (e.g., 3′-CTGG or CTGGT) is reached, gene 4 protein assembles a complementary tetranucleotide primer for lagging strand synthesis by T7 DNA polymerase.

In vivo as with purified proteins, either the 5′→3′ exonuclease activity of *E. coli* DNA polymerase I or the T7 gene 6 exonuclease can remove the T7 RNA primers.[45] Gaps are filled by T7 DNA polymerase; host or T7 ligase then joins the nascent fragments. T7 DNA polymerase prepared in the absence of EDTA cannot participate in the gap filling and joining reactions due to its ability to carry out strand displacement. The RNA-linked DNA chains isolated in vivo from T7-infected cells bear RNA primers that are mostly tetranucleotides of the sequence $pppACN_1N_2$ in which $N_1$ and $N_2$ are rich in C and A; pentanucleotides are present at a low frequency.[46] Furthermore, mapping of the in vivo Okazaki fragment initiation sites to a region of T7 DNA of known sequence reveals that the in vivo recognition sites for the gene 4 protein are 3′-CTGGN-5′ and 3′-CTGTN-5′.[47]

(ii) <u>Initiation of DNA replication at the T7 replication origin</u> requires T7 RNA polymerase. T7 DNA synthesis in vivo ceases after inactivation of the enzyme.[48] Two tandem T7 RNA polymerase promoters at the primary origin suggest a transcriptional role for T7 RNA polymerase in initiation of T7 DNA replication. Purified T7 RNA polymerase in conjunction with T7 DNA polymerase, T7 gene 4 protein, and the ribo- and deoxyribonucleoside triphosphates initiates DNA replication at the primary origin of an intact T7 DNA molecule.[49] The initiation event, scored by a replication bubble (eye), is site-specific; if the primary origin is deleted no initiation occurs. DNA synthesis is specific for recombinant plasmids containing the primary origin, provided the plasmid is converted to a linear form. However, unlike the bidirectional initiation in vivo, initiation in vitro is unidirectional and moves rightward.[50] The mechanism by which

44. Tabor, S., Engler, M. J., Fuller, C. W., Lechner, R. L., Matson, S. W., Romano, L. J., Saito, H., Tamanoi, F., and Richardson, C. C. (1981) *ICN–UCLA Symposia 22*, 387.
45. Tabor, S., Engler, M. J., Fuller, C. W., Lechner, R. L., Matson, S. W., Romano, L. J., Saito, H., Tamanoi, F., and Richardson, C. C. (1981) *ICN–UCLA Symposia 22*, 387.
46. Seki, T., and Okazaki, T. (1979) *N. A. Res. 7*, 1603; Ogawa, T., and Okazaki, T. (1979) *N. A. Res. 7*, 1621.
47. Fujiyama, A., Kohara, Y., and Okazaki, T. (1981) *PNAS 78*, 903.
48. Hinkle, D. C. (1980) *J. Virol. 34*, 136.
49. Romano, L. J., Tamanoi, F., and Richardson, C. C. (1981) *PNAS 78*, 4107.
50. Romano, L. J., Tamanoi, F., and Richardson, C. C. (1981) *PNAS 78*, 4107.

transcription initiates replication at the primary origin is not yet known. The transcript itself could serve as a primer for T7 DNA polymerase or, alternatively, transcription through the AT-rich region of the origin could expose the single gene 4 protein recognition site, thereby allowing synthesis of a tetraribonucleotide to prime rightward DNA synthesis.

### T5 Phage

In the 2-step entry of T5 DNA, host-cell metabolic energy is not required to draw the 3-$\mu$m linear molecule into the cell.[51] One of the two T5-encoded, early proteins (A2) required for the second step of the transfer process is a dimeric DNA-binding protein,[52] which may function with the other (A1) at the cell membrane to facilitate DNA entry.

Transcription and replication in T5-infected cells may be integrated by the formation of a single complex of proteins involved in these reactions (Section S5-6).

## S14-6  Large-Size Phages: T4 and Other T-Even (T2, T6)

### Virion

A detailed phage T4 restriction map[53] based on a novel two-dimensional electrophoretic separation has improved alignments with genetic markers to produce a superior physical genomic map (Fig. S14-3). Among the T4 genes involved in DNA replication (Tables 14-10, 14-12), several aspects of identity and function have been clarified as follows:

| Gene | Function |
| --- | --- |
| 41 | DNA helicase (GTPase and ATPase)[54] |
| 46, 47 | selective DNA degradation[55] |
| 39, 52, 60 | fork initiation, type II DNA topoisomerase[56] |
| (58) | (allelic with gene 61) |
| 61 | primase (probably) in RNA primer synthesis[57] |

51. Maltouf, A. F., and Labedan, B., personal communication.
52. Snyder, C. E., Jr., and Benzinger, R. H. (1981) J. Virol. 40, 248.
53. O'Farrell, P. H., Kutter, E., and Nakanishi, M. (1980) MGG 179, 421.
54. Liu, C.-C., and Alberts, B. M. (1981) JBC 256, 2813.
55. Mickelson, C., and Wiberg, J. S. (1981) J. Virol. 40, 65.
56. Liu, L. F., Liu, C.-C., and Alberts, B. M. (1979) Nat. 281, 456; (1980) Cell 19, 697.
57. Liu, C.-C., and Alberts, B. M. (1981) JBC 256, 2821.

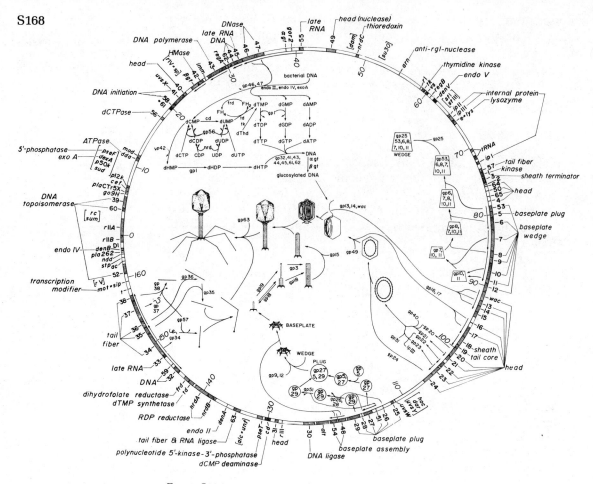

FIGURE S14-3
Genetic map of phage T4 with pathways of DNA synthesis and morphogenesis. (Courtesy of Professor E. Kutter.)

## Infection

Shutoff of host macromolecular syntheses within a minute after adsorption does not appear to be the result of a single dramatic event but rather of a chain of reactions with a multiplicative effect. In ordering the importance of these reactions the precedence of T4 over host promoters may prove to be of greater significance than putative nucleoid and membrane rearrangements.

Several origins for replication are likely—one near gene 41,[59] another near gene 5,[60] and others as well. The gene 39, 52, and 60 proteins, acting together as a type II topoisomerase[61] (Section S9-12), appear to be implicated in initiation of replication; some of the preferred sites of cleavage by the topoisomerase are located near replication origins.[62] The topoisomerase contains only phage-encoded proteins; host gyrase is a poor substitute when the T4 enzyme is defective.

Two alternative mechanisms are used by T4 to initiate DNA chains,[63] one dependent on the host RNA polymerase and the other on general recombination. Continued (late) DNA synthesis is sensitive to rifampicin under recombination-defective conditions but remains vigorous in the presence of the inhibitor when recombination generates D-loop structures.

Seven proteins are responsible for advancing the replication fork, and all are T4-encoded and have been purified to homogeneity. Gene 41 protein is a helicase and works with gene 61 protein, likely the primase, in vitro[64] to produce a pentaribonucleotide, identified as pppApCpNpNpN, that corresponds to the 5′ termini of nascent DNA fragments produced in vivo.[65] Gene 43 protein, the polymerase, achieves high processivity and fidelity,[66] with rates approaching the 500-nucleotides/sec value attained in vivo[67] through the conjoined actions of gene 44, 62, and 45 proteins,[68] the "sliding clamp" for the polymerase on the template. Binding of the single-stranded regions at the fork by gene 32 protein (Section S9-3) is essential for optimal advancement of the replication fork.

Complexes and membranes appear to physically link T4 DNA and the induced enzymes of deoxynucleotide biosynthesis with the replication apparatus influencing its operation as well as channeling the limiting deoxynucleotides to the site of DNA synthesis (p. 537). Evidence that T4 ribonucleoside diphosphate reductase, topoisomer-

58. Alberts, B. M., Barry, J., Bedinger, P., Burke, R. L., Hibner, U., Liu, C.-C., and Sheridan, R. (1980) ICN–UCLA Symposia 19, 449.
59. Mosig, G., Luder, A., Rowen, L., Macdonald, P., and Bock, S. (1981) ICN–UCLA Symposia 22, 277; Dannenberg, R., and Mosig, G. (1981) J. Virol. 40, 890.
60. Halpern, M. E., Mattson, T., and Kozinski, A. W. (1979) PNAS 76, 6137; Kozinski, A. W., Ling, S.-K., Hutchinson, N., Halpern, M. E., and Mattson, T. (1980) PNAS 77, 5064.
61. Liu, L. F., Liu, C.-C., and Alberts, B. M. (1979) Nat. 281, 456; (1980) Cell 19, 697.
62. Kreuzer, K., and Alberts, B. M., personal communication.
63. Luder, A., and Mosig, G. (1982) PNAS 79, 1101.
64. Liu, C.-C., and Alberts, B. M. (1981) JBC 256, 2813; Nossal, N. G. (1980) JBC 255, 2176; Liu, C.-C., and Alberts, B. M. (1980) PNAS 77, 5698; (1981) JBC 256, 2821.
65. Kurosawa, Y., and Okazaki, T. (1979) JMB 135, 841.
66. Hibner, U., and Alberts, B. M. (1980) Nat. 285, 300; Sinha, N. (1981) JBC 256, 10671.
67. Alberts, B. M., Barry, J., Bedinger, P., Burke, R. L., Hibner, U., Liu, C.-C., and Sheridan, R. (1980) ICN–UCLA Symposia 19, 449.
68. Huang, C.-C., Hearst, J. E., and Alberts, B. M. (1981) JBC 256, 4087.

ase, and T4 DNA play additional interactive roles in vivo in deoxynucleotide synthesis[69] is a further incentive for isolation and characterization of this remarkable membrane-bound, replisomal complex.

## S14-7  Temperate Phages: λ, P22, P2, P4, P1, Mu

Phage λ[70]

The early replication of phage λ, as for *E. coli*, is bidirectional and symmetric from a unique origin. "Transcriptional activation" by RNA polymerase (p. 546), thought to alter the structure of the origin or its cellular location, can be spatially uncoupled by at least 95 bp from the origin with the transcription directed away from it.[71]

Later in infection an alternative asymmetric rolling-circle mode of replication generates the concatemeric structures that are processed and packaged into virions. As in the control of gene expression, studies of phage λ regulation should provide insights into how bidirectional replication is initiated and how replication is either switched to a rolling-circle mechanism in lytic development or switched off in the lysogenic response.

*O and P proteins.*  The major recent progress has been the isolation and characterization of the O and P proteins responsible for initiation of bidirectional replication. The 36-kdal O polypeptide, with an in vivo half-life of only 1.5 minutes because of ATP-dependent proteolysis, can be stabilized by metabolic poisons (e.g., cyanide, arsenate, or azide) that rapidly deplete the cellular ATP pool.[72] Cloning in a chimeric plasmid also leads to massive accumulation of O protein with levels ($10^5$ dimers per cell) reaching 10 percent of the soluble protein.[73] The 26-kdal P polypeptide, assayed by complex formation with dnaB protein,[74] can also be overproduced 10- to 25-fold in cells with thermoinducible plasmids containing the P gene.[75]

The properties of purified O protein are particularly intriguing.[76]

69. Wirak, D. O., and Greenberg, G. R. (1980) *JBC 255*, 1896; Chiu, C.-S., Cox, S. M., and Greenberg, G. R. (1980) *JBC 255*, 2747; Greenberg, G. R., personal communication.
70. *Lambda II* (successor to *Bacteriophage Lambda*, 1971) (1982), CSHL, in press.
71. Furth, M. E., Dove, W. F., and Meyer, B. J. (1982) *JMB 154*, 65.
72. McMacken, R., and Roberts, J. D., personal communication.
73. McMacken, R., and Roberts, J. D., personal communication.
74. Klein, A., Lanka, E., and Schuster, H. (1980) *EJB 105*, 1.
75. Mallory, J., and McMacken, R., personal communication.
76. Tsurimoto, T., and Matsubara, K. (1981) *N. A. Res. 9*, 1789.

The protein binds specifically to *ori* DNA; a maximal segment of about 95 base pairs is protected against DNase attack. This segment contains four tandem, nearly identical 19-bp sequences. At a low O protein concentration, mainly the inner two sequences are protected; at higher O protein levels, all four sequences are protected. Each 19-bp repeat has a region of dyad symmetry. Were the dyad symmetry responsible for dimer binding as in the case of cro and cI (Section S9-1), then the *ori* region would bind up to eight monomers.

The binding properties of O protein signify a sensitive regulatory system for the origin of replication, and possibly for determining the "on-off" state of replication and controlling the switch from the symmetric to asymmetric mode of replication. The high levels of O protein found early in infection might signal bidirectional starts, and the low levels late in infection, after repression by cro has set in, might initiate asymmetric replication. It is significant that $cro^-$ mutants are defective in λ replication at late times.[77] By analogy with the host oriC system (p. 545), the O and P proteins are the respective counterparts of *E. coli* dnaA and dnaC proteins. Interaction of O protein with P protein links them to the host dnaB protein and the complex *E. coli* replication system on which λ depends.

Homogeneous O and P proteins added to an oriC enzyme system (Section S11-15) have been shown to initiate vigorous replication from a λ origin.[77a]

## Other Temperate Phages

A composite physical and genetic map of phage P22 and further mapping data on phage P4 are available.[78] (In correction of Table 14-13, p. 540, the phage P4 DNA is 11.2 kb; it is a satellite of phage P2 and lysogenizes *E. coli* from a defined attachment site;[79] the remark about plasmid state and low copy number applies to phage P1 on the next line.) P4-encoded α protein is responsible for two activities: (i) an RNA polymerase that synthesizes poly rG from a poly dG·poly dC template, and (ii) an activity, still undefined, essential for in vitro replication of P4 DNA specifically in the presence of host DNA polymerase and single-strand binding protein.[80]

The *gin* gene product of phage Mu responsible for site-specific, intramolecular recombination is a 21.5-kdal protein.[81]

77. Folkmanis, A., Maltzman, W., Mellon, P., Skalka, A., and Echols, H. (1977) *Virol.* 81, 352.
77a. McMacken, R., personal communication.
78. Rutila, J. E., and Jackson, E. N. (1981) *Virol.* 113, 769; Kahn, M., Ow, D., Sauer, B., Rabinowitz, A., and Calendar, R. (1980) *MGG* 177, 399.
79. Calendar, R., Ljungquist, E., Deho, G., Usher, D. C., Goldstein, R., Youderian, P., Sironi, G., and Six, E. W. (1981) *Virol.* 113, 20.
80. Krevolin, M. (1981) *Studies on DNA Replication of Phages P2 and P4.* Ph.D. thesis, Univ. of Calif., Berkeley.
81. Kwoh, D., and Zipser, D. (1981) *Virol.* 114, 291.

## S14-8 Other Phages: N4, PM2, and *Bacillus* Phages (SPO1, PBS1, PBS2, φ29, SPP1)

### N4 Phage

N4 phage, a coliphage with a duplex 45-Mdal genome, is the only example of a bacterial virion that contains its own transcriptional enzyme. In the host cell, the virion RNA polymerase (Section S7-8) directs rifampicin-resistant transcription of N4 early genes, including that of the RNA polymerase for late transcription. The N4 virion RNA polymerase has the novel feature, shared only by the vaccinia polymerase, of DNA-dependent ATPase activity.[82] The structure of the enzyme is still in doubt (p. 554). Host DNA synthesis is blocked after infection but DNA is not degraded; RNA and protein syntheses are maintained. N4 induces a 94-kdal DNA polymerase[83] which, like T4 DNA polymerase, prefers a single-stranded DNA template and depends on single-strand binding protein.

### φ29 Phage

Gene 3 protein is attached covalently through serine to the phosphoryl group of the 5′-terminal residue (dAMP).[84] A possible role of the protein in priming initiation of replication of this duplex genome from each end is comparable to that of the terminal adenoviral protein (Section S15-4).[85] Aphidicolin inhibition of replication of φ29 and related phage M2, but not of other *B. subtilis* phages (e.g., SP50, SPO1, SPP1) or host DNA[86] (Section S12-5), suggests some specific interaction of the drug with the 5′-phosphoryl-linked protein. Phage assembly in vitro (180 particles per cell equivalent) approaches in vivo efficiency and matches that of other packaging systems (e.g., phages λ, T3, T7, T4, P2, P22).[87] Proheads, gene 16 protein, and ATP participate in packaging the gene 3 protein-DNA complex without involvement of concatemeric DNA intermediates.

### PBS2 Phage

The uracil-DNA glycosylase (Section S16-2) inhibitor encoded by this uracil-containing *B. subtilis* phage is effective against uracil-DNA glycosylases from other cells (e.g., *E. coli,* yeast, human) but is inactive against glycosylases specific for other base residues in DNA.[88]

82. Rothman-Denes, L. B., personal communication.
83. Sugino, A., Rist, J. K., and Rothman-Denes, L. B., personal communication.
84. Mellado, R. P., Penalva, M. A., Inciarte, M. R., and Salas, M. (1980) *Virol. 104,* 84; Harding, N. E., and Ito, J. (1980) *Virol. 104,* 323; Hermoso, J. M., and Salas, M. (1980) *PNAS 77,* 6425.
85. Sogo, J. M., Garcia, J. A., Peñalva, M. A., and Salas, M. (1982) *Virol. 116,* 1.
86. Yoshikawa, H., and Ito, J. (1981) *PNAS 78,* 2596.
87. Bjornsti, M.-A., Reilly, B. E., and Anderson, D. L. (1981) *PNAS 78,* 5861.
88. Karran, P., Cone, R., and Friedberg, E. C. (1981) *B. 20,* 6092.

The importance of plasmids, the enormous surge of research interest in them, and the significant advances in understanding the mechanisms and control of their replication would elevate their place in a revised edition of *DNA Replication* from sectional status to a large chapter. These recent developments stem largely from the use of recombinant DNA technology, discoveries of in vitro replication systems for plasmids, and the availability of rapid nucleotide sequencing techniques. Several gene cloning strategies have been of special value:

(i)   Construction of DNA fragments that lack replication functions but carry selective markers (e.g., antibiotic resistance) to select DNA fragments that specify replication functions.

(ii)   Separation of replication determinants on two different and compatible plasmids to facilitate the identification of *trans*-acting replication functions, positive acting for plasmids R6K and RK2 and negative acting for R1 and λdv (see below).

(iii)   Fusion of two replicons (plasmid or phage), one being subjected to mutations and the other being conditionally defective (e.g., ColE1 replication in *polA*+ vs. *polA*− cells). In this way, replication-defective mutants of a plasmid can be maintained under the permissive condition that permits replication of the conditional replicon; the defect in replication can be analyzed under the restrictive condition that does not permit replication of the conditional replicon.

Control of Plasmid Copy Number

Some novel insights,[89] as well as a firmer basis for earlier indications (p. 464),[90] provide a clearer picture of the molecular details of control mechanisms.

(i)   The nucleotide sequence of the replication region of ColE1[91] and complete sequences of R6K[92] and R1[93] are in hand.

(ii)   Regulation is at the level of initiation.

89.  Kolter, R. (1981) *Plasmid 5*, 2; Timmis, K. N., Danbara, H., Brady, G., and Lurz, R. (1981) *Plasmid 5*, 53; Matsubara, K. (1981) *Plasmid 5*, 32; Pritchard, R., and Grover, N. (1981) in *Mol. Biol. Plasmids*, p. 271.

90.  Kolter, R., and Helinski, D. R. (1979) *Ann. Rev. Gen. 13*, 355; Tomizawa, J., and Selzer, G. (1979) *ARB 48*, 999.

91.  Oka, A., Nomura, N., Morita, M., Sugisaki, H., Sugimoto, K., and Takanami, M. (1979) *MGG 172*, 151.

92.  Stalker, D. M., Shafferman, A., Tolun, A., Kolter, R., Yang, S., and Helinski, D. R. (1981) *ICN–UCLA Symposia 22*, 113.

93.  Oertel, W., Kollek, R., Beck, E., and Goebel, W. (1979) *MGG 171*, 277; Rosen, J., Ohtsubo, H., and Ohtsubo, E. (1979) *MGG 171*, 287; Stougaard, P., Molin, S., Nordstrom, K., and Hansen, F. G. (1981) *MGG 181*, 116; Armstrong, K., Rosen, J., Ryder, T., Ohtsubo, E., and Ohtsubo, H. (1981) in *Mol. Biol. Plasmids*, p. 279.

(iii)  Rate of initiation is determined by concentration of plasmid-specified inhibitors: an RNA and protein for ColE1 and R1; a protein (*tof* repressor) for λdv.

Emergence of a small RNA as the regulatory device for initiation of DNA synthesis at the origin of plasmid replication is surprising but not entirely without precedent. Generally, specific recognition for action at a DNA site is the preserve of a protein. Yet the specificity of RNA recognition of DNA or its RNA complement can be great, and a regulatory role was anticipated in the use of RNA secondary structure in attenuation of transcription (Section S7-6).

(a) *ColE1 and related plasmids (p15A, RSF1030, and CloDF13).* Mutant analyses indicated[94] and in vitro replication studies demonstrated[95] that an RNA transcript 108 nucleotides long (called RNA I) inhibits RNA priming of the DNA start (see below). RNA I is transcribed from the strand opposite to the template for the primer, 400 to 480 bp upstream from the origin. RNA I, by hybridizing to DNA or RNA, interferes in some undisclosed manner with the hybridization of the RNA primer with its template.[96] The primer-template substrate, required for RNaseH action, is essential for initiation of DNA synthesis.

Mutant plasmids with single base changes in the RNA I region have altered properties in vivo and in the inhibition of primer formation in vitro. Based on these analyses, RNA I qualifies as an important element in the regulation of compatibility between plasmids (see below) and the control of copy number. It seems likely too that a region downstream from the origin encodes a small protein that also limits plasmid copy number.[97]

(b) *R1 and related plasmids (R6 and R100).*[98,99] Replication of this plasmid is controlled by the products of two genes, *copA* and *copB*, that act as inhibitors of replication. A small RNA (80–90 residues) synthesized from the *copA* region and an 11-kdal polypeptide encoded by *copB* are the gene products responsible for copy number control and compatibility. Instability of the *copA* RNA in cells suggests that cyclic changes in its concentration could determine the initiation frequency of replication.

94. Conrad, S. E., and Campbell, J. L. (1979) *Cell 18*, 61; Muesing, M., Tamm, J., Shepard, H. M., and Pollsky, B. (1981) *Cell 24*, 235; Stuitje, A. R., Spelt, C. E., Veltkamp, E., and Nijkamp, H. J. J. (1981) *Nat. 290*, 264.

95. Itoh, T., and Tomizawa, J. (1980) *PNAS 77*, 2450; Tomizawa, J., Itoh, T., Selzer, G., and Som, T. (1981) *PNAS 78*, 1421; Tomizawa, J., and Itoh, T. (1981) *PNAS 78*, 6096.

96. Lacatena, R. M., and Cesareni, G. (1981) *Nat. 294*, 623; Hillenbrand, G., and Staudenbauer, W. L. (1982) *N. A. Res. 10*, 833.

97. Twigg, A. J., and Sherratt, D. (1980) *Nat. 283*, 216.

98. Molin, S., Stougaard, P., and Nordström, K. (1981) *Microbiology 1981*, p. 408, ASM, Washington, D.C.; Danbara, H., Brady, G., Timmis, J. K., and Timmis, K. N. (1981) *PNAS 78*, 4699; Burger, K. J., Steinbauer, J., Röllich, G., Kollek, R., and Goebel, W. (1981) *MGG 182*, 44; Rownd, R., Easton, A., and Sampathkumar, P. (1981) in *Mol. Biol. Plasmids*, p. 303.

99. Stougaard, P., Molin, S., and Nordström, K. (1981) *PNAS 78*, 6008.

(c) λdv.[100]   The tof protein repressor appears to have the central role in regulating replication of the plasmid, although a second level of control in the region of the origin may also exist.

(iv)   The site of control of initiation of plasmid replication is generally remote from the origin of replication: 0.5 kb for ColE1, 1.2 kb for R1, 2 kb for R6K,[101] and 1.1 kb for λdv. Deletion of a short segment or even a base-pair substitution between the control and origin sites may block initiation.

(v)   Direct repeats of nucleotide sequences have been identified in the region of replication initiation and control for several plasmids (R6K, RK2, mini-F, λdv) but not in others (R1 and ColE1). R6K has seven 22-bp repeats in tandem array in one of its three origins.[102] These repeats express R6K incompatibility when cloned into plasmid pBR322.[103] Direct repeats have also been identified in the origin region of RK2 (eight 17-bp repeats)[104] and in an incompatibility region of mini-F (five 22-bp repeats).[105] In the case of mini-F, two of the five 22-bp direct repeats have been cloned as a 58-bp fragment and shown to express strong incompatibility against Flac and various F plasmid derivatives.

(vi)   The replication terminus of plasmid R6K when cloned into ColE1-type replicons functions both in vivo[106] and in vitro.[107] A nucleotide sequence of 215 bp containing this replication terminus lacks any twofold rotational symmetry.[108]

(vii)   Replication of multicopy plasmids is independent of the host chromosome cycle. Choice of a plasmid for replication is random; initiation of a newly replicated plasmid is as likely as that of any other.

(viii)   Incompatibility of two different plasmids, defined as the maintenance of one to the exclusion of the other after several cell generations, is based on:

(a) inability of the copy number control mechanism to distinguish between the two plasmids,

(b) randomness in the choice for replication among all plasmids in the cellular pool, and

(c) relatively random partitioning of plasmids between daughter cells at cell division.

100. Matsubara, K. (1981) Plasmid 5, 32.
101. Shafferman, A., Stalker, D., Tolun, A., Kolter, R., and Helinski, D. R. (1981) in Mol. Biol. Plasmids, p. 259.
102. Stalker, D., Kolter, R., and Helinski, D. R. (1979) PNAS 76, 1150.
103. Shafferman, A., Stalker, D. M., Tolun, A., Kolter, R., and Helinski, D. R. (1981) in Mol. Biol. Plasmids, p. 259.
104. Stalker, D. M., Thomas, C. M., and Helinski, D. R. (1981) MGG 181, 8.
105. Tolun, A., and Helinski, D. R. (1981) Cell 24, 687.
106. Kolter, R., and Helinski, D. R. (1978) JMB 124, 425.
107. Germino, J., and Bastia, D. (1981) Cell 23, 681.
108. Bastia, D., Germino, J., Crosa, J. H., and Ram, J. (1981) PNAS 78, 2095.

(ix) Low-copy-number plasmids are rarely lost by a cell. This plasmid maintenance in all cells of a population implies a cellular partition mechanism that recognizes a special plasmid region. For lack of such a region in derivatives of pSC101,[109] R1,[110] and R100,[111] the plasmid disappears from a growing population of cells at the frequency calculated for random assortment. A 270-bp region, designated *par* in pSC101 and adjacent to the region for replication initiation and control, is presumed to be responsible for active segregation and stabilization in the cell population. Such stabilizing elements are also found in R1 and R100 (called *stb*) and are transferable from one type of plasmid to another.

## In Vitro Plasmid Replication Systems

*ColE1*.[112] Primer formation by RNA polymerase transcription is initiated 555 bp upstream from the origin (Fig. S14-4). To be utilized by DNA polymerase I to start the leading L strand (from the origin), the RNA primer must be partially degraded by RNaseH. Earlier indications of a role for RNaseIII (p. 564) have not been confirmed. The failure of DNA polymerase III holoenzyme to substitute for DNA polymerase I in this first stage of replication is likely due to its lack of an effective 5'→3' exonuclease activity to displace and eliminate primer fragments left after RNaseH action. For lack of this activity, the DNA polymerase I large fragment (p. 139) also fails to extend the ColE1 primer by even a single nucleotide.[113] Details regarding ColE1 replication beyond synthesis of the initial 6S DNA fragment are lacking except for the likely participation of DNA polymerase III and the primosome system.[114] A primosome assembly site and DNA chain start on the L strand can be associated with discontinuous synthesis of the lagging strand; an assembly site and chain start at a site on the H strand downstream from the origin may be the start of the conjugal transfer strand (Section S11-9).

*R1*. An effective in vitro system for replication of exogenously added R1 DNA as well as the endogenous plasmid DNA in extracts of cells harboring R1 requires an active, coupled, transcription-translation system.[115] Inhibition of transcription (e.g., rifampicin) re-

109. Meacock, P. A., and Cohen, S. N. (1980) *Cell 20*, 529.
110. Nordström, K., Molin, S., and Aagard-Hansen, H. (1980) *Plasmid 4*, 332, (1981) in *Mol. Biol. Plasmids*, p. 291.
111. Miki, T., Easton, A. M., and Rownd, R. H. (1980) *J. Bact. 141*, 87.
112. Itoh, T., and Tomizawa, J. (1980) *PNAS 77*, 2450; Tomizawa, J., Itoh, T., Selzer, G. and Som, T. (1981) *PNAS 78*, 1421; Tomizawa, J., and Itoh, T. (1981) *PNAS 78*, 6096; Staudenbauer, W. L., Scherzinger, E., and Lanka, E. (1979) *MGG 177*, 113; Hillenbrand, G., and Staudenbauer, W. L. (1982) *N. A. Res. 10*, 833.
113. Tomizawa, J., personal communication.
114. Staudenbauer, W. L., Scherzinger, E., and Lanka, E. (1979) *MGG 177*, 113.
115. Diaz, R., Nordström, K., and Staudenbauer, W. L. (1981) *Nat. 289*, 326; Khan, S. A., Carleton, S. M., and Novick, R. P. (1981) *PNAS 78*, 4902.

Primer formation

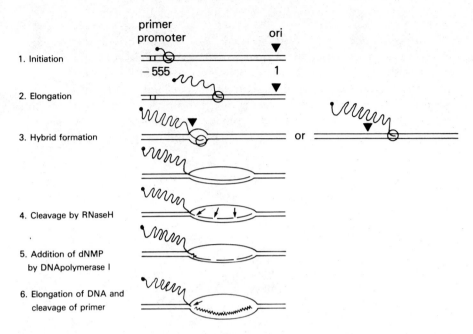

1. Initiation

2. Elongation

3. Hybrid formation

4. Cleavage by RNaseH

5. Addition of dNMP
   by DNApolymerase I

6. Elongation of DNA and
   cleavage of primer

FIGURE S14-4
Proposed mechanism of primer formation in the initiation of DNA synthesis of plasmid ColE1. (Courtesy of Dr. J. Tomizawa.)

duces replication by 85 percent and inhibitors of translation (e.g., puromycin or chloroamphenicol) reduce replication by 99 percent, presumably by interrupting the production of an essential plasmid-encoded protein whose level or stability is inadequate in the extracts of plasmid-bearing cells.

In vitro replication of the small pSC101 plasmid (9 kb) has been observed[116] and shows requirements similar to those described for R1. Added interest attaches to the enzymatic resolution and reconstitution of pSC101 replication because of in vitro[116] and in vivo[117] evidence that it depends on *dnaA* function.

*R6K.* This 38-kb plasmid (p. 566) possesses three distinctive origins, all of which function in vivo[118] and in vitro.[119] The 35-kdal π protein, the sole plasmid-encoded protein, is required for initiation

116. Kaguni, J. M., Fuller, R., Flynn, J., and Kornberg, A., unpublished observation.
117. Hasunuma, K., and Sekiguchi, M. (1979) *J. Bact.* 137, 1095; Frey, J., Chandler, M., and Caro, L. (1979) *MGG*, 174, 117; Felton, J., and Wright, A. (1979) *MGG* 175, 231.
118. Inuzuka, N., Inuzuka, M., and Helinski, D. R. (1980) *JBC* 255, 11071.
119. Crosa, J. H. (1980) *JBC* 255, 11075.

of replication from each of these origins. It does not function as a positive regulatory element[120] and its role is still unclear.

*Staphylococcus aureus plasmid pT181.*   An effective cell-free system is specific for replication of pT181 and closely related plasmids, providing extracts are prepared from cells that can supply the required plasmid-encoded substance.[121] The lack of rifampicin inhibition indicates that in vitro transcription and translation are not needed to provide the levels of plasmid-encoded factor required for replication of exogenous as well as endogenous plasmid DNA.

*OriC plasmids.*   The potent enzyme system for replication of plasmids that depend on the cloned origin of the *E. coli* chromosome is described in detail in Section S11-15.

Conjugal Transfer[122]

A large fraction of a conjugative plasmid is devoted to conjugal transfer genes. For an F plasmid it covers about 20 genes or about one-third of the 100-kb molecule; a comparable fraction of the small ColE1 molecule is concerned with mobilization for transfer. For a replicon to be transferred, there are three basic requirements:
   (i)   an origin of transfer (*oriT*) sequence,
   (ii)   synthesis by the cell of a DNA transfer and replication (*Dtr*) system that specifically recognizes *oriT*,
   (iii)   synthesis by the cell of a system for stable mating pair formation (*Mpf*) with recipient cells that specifically recognize this *Dtr* system, and
   (iv)   encoding of the *Dtr* and *Mpf* systems by the replicon or provision of them in *trans* by other replicons.

The roles of some of the many genes in the *Dtr* and *Mpf* systems have been identified. However the mechanisms for forming and using the "relaxation complexes" (p. 566) that mark the *oriT* sites in ColE1 and F plasmids remain unknown.

Transposons

The subject of transposition is covered under "recombination" (Section S16-4). The recombinational events that accomplish the transposition of movable elements are in some cases linked to a duplication of the element. Many schemes to accomplish this replicative fusion have been suggested, but detail is lacking in all of them.

120. Shafferman, A., Stalker, D., Tolun, A., Kolter, R., and Helinski, D. R. (1981) in *Mol. Biol. Plasmids*, p. 259.
121. Khan, S. A., Carleton, S. M., and Novick, R. P. (1981) *PNAS 78*, 4902.
122. Willetts, N. (1981) in *Mol. Biol. Plasmids*, p. 207.

# S15

# Animal DNA Viruses, Retroviruses, and Organelles

## Abstract

Sequence determinations of viral genomes have provided clues to the mechanism of replication but are still inadequate for delineating the enzymological process. Only for the adenoviruses, retroviruses, and yeast plasmids have in vitro systems been obtained that are suitable for purification and molecular characterization. For the duplex circular papovaviruses (e.g., SV40, polyoma), T antigen is known to be essential but an in vitro system for initiating replication has not yet been developed. The single-stranded linear genome of parvoviruses likely depends entirely on host enzymes for replication and on the extension of terminal hairpins to prime the process. Integration of the parvoviral genome into the host chromosome also uses these unique termini and may be the best model for studies of such recombinational events in eukaryotes.

Best understood of the viral replicative mechanisms is that of the linear duplex adenovirus in which a virus-encoded protein appears to start synthesis by covalent addition of the 5′-terminal nucleotide. If proven with purified replicative enzymes, this mechanism of initiation would be the sole instance in which RNA priming is not used to initiate a DNA chain.

Despite detailed analyses of the multiple arrangements of a herpes virus genome, little is known about the replicative activities of the viral polymerase and other induced proteins, and even less is known about how the genome is integrated. The large pox viruses are replicated in cytosolic virosome factories staffed with viral enzymes (including a characterized polymerase), but no in vitro system for replication has been established. Retroviral RNA integration through stages of reverse transcription and replication that form duplex DNA are known only in outline, and the final stages are especially vague.

With mammalian mitochondrial DNA, available information about (i) its sequence, (ii) the origins of replication, (iii) the nature of the DNA polymerase, and (iv) the initial D-loop intermediates, has not been tied together with enzymological details. The most attractive system for probing the nature of the replication of the cellular genome is provided by yeast plasmids that rely on cell-development-cycle genes and controls. Purification of the in vitro systems that initiate and elongate these duplex circles offers the same promise for exploring eukaryotic replication that oriC plasmids do for prokaryotic genomes.

## S15-2 Papovaviruses: Simian Virus 40 (SV40) and Polyoma

### Genome

The entire nucleotide sequences for the DNAs from the three papovaviruses, SV40, polyoma, and BKV have been determined.[1] The origin of DNA replication in the three viruses is in a highly conserved region of about 40 base pairs within a noncoding region of about 450 base pairs, which also contains the promoters for early and late mRNAs. All regulatory signals for replication and transcription are thus located within a short stretch of the genome. Numerous mutations in and around the origin region have revealed its dimensions and functional aspects. In some cases the effect of a mutation has been suppressed by a second-site mutation in the large T antigen.[2]

### Large T Antigen[3]

This 96-kdal protein is now readily purified[4] and is phosphorylated at certain serine and threonine residues. As an isologous tetramer, T antigen recognizes and binds specifically to the SV40 DNA region

1. Tooze, J. (1980) in *Molecular Biology of Tumor Viruses*, part 2, 2nd ed. (Tooze, J., ed.), CSHL.
2. Shortle, D. R., Margolskee, R. F., and Nathans, D. (1979) *PNAS 76*, 6128.
3. Tjian, R. (1981) *Cell 26*, 1; *Curr. Top. 93*, 5; Tooze, J. (1980) in *Molecular Biology of Tumor Viruses*, part 2, 2nd ed. (Tooze, J., ed.), CSHL.
4. Tegtmeyer, P., and Andersen, B. (1981) *Virol. 115*, 67.

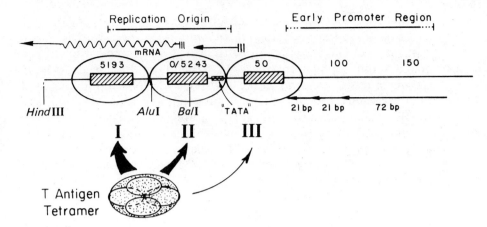

FIGURE S15-1
Model for cooperative binding of T antigen to SV40 DNA in viral DNA replication
and the control of early transcription. In the region of the SV40 genome containing
the three T antigen binding sites (boxes I, II, and III), numbers above the boxes
are the nucleotide number of the center of each binding site, with the *BgI*I restriction
site as zero. Horizontal arrows depict 21-bp and 72-bp direct repeated sequences.
Bars indicate regions required for efficient viral DNA replication in vivo and early
transcription in vitro. At early times after viral infection, when the binding sites are
free of T antigen, RNA polymerase II can initiate transcription at the early promoter
region. A tetramer of T antigen binds at site I and coordinates binding of a second
tetramer at site II, allowing DNA replication to begin. As the infection progresses,
the T antigen concentration increases to a level that allows cooperative binding
of a third tetramer to site III. At this level of T antigen all classes of early transcripts
would be repressed. (Courtesy of Professor R. Tjian.)

that contains the replication origin and the promoter for early
transcription[5] (Fig. S15-1). Binding is not affected by dephosphoryla-
tion of the serine residues representing the majority of the phosphorylated sites.[6] T antigen also has an intrinsic DNA-dependent
ATPase activity with no functional significance known as yet. In
binding supercoiled SV40 DNA, unwinding activity, like that observed with topoisomerases or RNA polymerase, was not detected,
nor is the binding to the viral origin affected by the topological
state of the DNA.[7] Presumably T antigen functions as a signal or
recognition subunit of larger assemblies that initiate replication and
transcription.

Replication

The precise starts for bidirectional replication are not known, and
for lack of an in vitro system that performs initiation the biochemis-

5. Myers, R. M., Rio, D. C., Robbins, A. K., and Tjian, R. (1981) *Cell* 25, 373; Tegtmeyer, P.,
   Andersen, B., Saw, S. B., and Wilson, V. G. (1981) *Virol.* 115, 75.
6. Shaw, S. B., and Tegtmeyer, P. (1981) *Virol.* 115, 88.
7. Myers, R. M., Kligman, M., and Tjian, R. (1981) *JBC* 256, 10156.

try of this key stage remains obscure. After injection of SV40 or polyoma DNA into unfertilized *Xenopus* eggs, the viral DNAs are replicated in phase with the cell cycle during maturation of the oocyte,[8] but replication does not depend on the origin region. Origins are also random in the low level of DNA replication that remains when a functional large T antigen is absent in infections at a restrictive temperature with a temperature-sensitive gene A (large T) mutant of SV40.[9] These results suggest that T antigen may represent a class of proteins that enables a virus to initiate many rounds of replication without restraint by the normal controls of cellular DNA replication.

### Transformation

In addition to large T and small T antigens, polyoma, unlike SV40, also encodes a third early protein, the middle t antigen. This protein merits special attention since it appears to act as a primary inducer of transformation[10] and also has connected to it a protein kinase activity, specifically phosphorylating tyrosine residues,[11] a mechanism that in other instances has been linked to transforming activity.

## S15-3   Parvoviruses: Nondefective and Defective[12]

The small, linear, single-stranded genomes of these viruses, depending almost entirely on host functions, remain attractive model substrates for eukaryotic DNA replication.

### Virions

Distinctions between the virions of the two classes of parvoviruses are fading. Autonomous (nondefective) parvoviruses had been described as containing only (−) strands, but there is now an example of one that resembles the defective adenoassociated viruses (AAV) in packaging (+) and (−) strands without preference.[13] The

---

8. Harland, R. M., and Laskey, R. A. (1980) *Cell 21*, 761.
9. Martin, R. G., and Stelow, V. P. (1980) *Cell 20*, 381.
10. Ito, Y., Spurr, N., and Griffin, B. E. (1980) *J. Virol. 35*, 219.
11. Eckhart, W., Hutchinson, M., and Hunter, T. (1979) *Cell 18*, 925.
12. Hauswirth, W. W., and Berns, K. I. (1981) in *Organization and Replication of Viral DNA* (Kaplan, S., ed.), CRC Press, Cleveland in press; Berns, K. I., and Hauswirth, W. W. (1979) *Adv. Virus Res. 25*, 407; Mitra, S. (1980) *Ann. Rev. Gen. 14*, 347.
13. Majaniemi, I., Tratschin, J. D., and Siegl, G. (1981) Fifth International Congress of Virology, Strasbourg, August 1981 (abstracts) p. 366; Bates, R. C., Bannerjee, P. T., and Mitra, S., personal communication.

two groups share other features: terminally hinged parental and progeny duplex molecules, and dual sequences for at least one of the termini.

Replication Mechanism for AAV

Direct determination of AAV terminal sequences confirms the predictions of _hairpin priming_ and _hairpin-loop transfers_.[14] The four possible combinations of terminal sequences for each strand are found in equal proportions, showing that there is no bias for a particular orientation of the terminal sequences (Fig. S15-2). Together with already existing evidence for terminally crosslinked molecules, both in parental and progeny intracellular replicative intermediates,[15] these results are in accord with terminal hairpin strand initiation, and a hairpin-loop-transfer mechanism to replicate 5'-terminal sequences.

Analysis of the specific activities in restriction fragments of pulse-labeled intracellular duplex replicative forms of AAV DNA indicates that synthesis of both strands takes place by continuous elongation in the 5'→3' direction by a strand-displacement mechanism similar to that proposed for adenovirus (Section S15-4). The displaced single strands are either encapsidated or recycled back to replicative form by synthesis of the complementary strand, as in the initial synthesis of parental replicative form from infecting viral single strands (Fig. S15-2).

AAV is integrated into cellular DNA in latently infected cells through junctions with the ends of the viral genome.[16] In the use of these unique viral sequences, integration of AAV differs from the nonspecific selection of SV40 sequences in its integration (p. 575).[17] Rescue of integrated AAV achieved by adenovirus superinfection depends on adenoviral tumor antigen(s) and might involve the AAV termini in a manner resembling their use in replication.

Replication Mechanism for Autonomous Viruses

Virtually the entire DNA sequence for some of the autonomous parvoviruses is now also known, but the replication mechanism is less well understood than for AAV. As in the case of AAV, the 5'-terminal sequence for autonomous parvoviruses exists in two

14. Lusby, E., Fife, K. H., and Berns, K. I. (1980) _J. Virol. 34,_ 402; Lusby, E., Bohenzky, R., and Berns, K. I. (1981) _J. Virol. 37,_ 1083.
15. Hauswirth, W. W., and Berns, K. I. (1979) _Virol. 93,_ 57.
16. Ostrove, J. M., Duckworth, D. H., and Berns, K. I. (1981) _Virol. 113,_ 521; Berns, K. I., personal communication.
17. Weinberg, R. A. (1980) _ARB 49,_ 197; Stringer, J. R. (1981) _J. Virol. 38,_ 671.

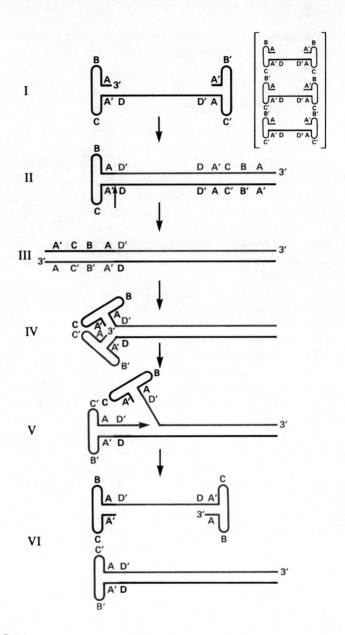

FIGURE S15-2

Proposed model for replication of AAV. I. Viral strand(s). The letters, A, A′, B, . . . refer to the terminal repeated sequences. At the left end, A represents nucleotides 1–41; B, 42–62; C, 64–84; A′, 85–125; and D, 126–145. For a particular strand, the terminal sequences are always inverted with respect to each other for A and D, but only for two out of the four possible combinations of B and C. Of the four possible terminal configurations for a strand, three are shown in brackets. II. A hairpin loop has primed synthesis of complementary strand to form duplex replicative form. (Note that the distance between C and B in the linear segment on the right should be the same as in the folded form on the left.) III. Nicking (arrow) opposite 3′

forms,[18] but only a single sequence has been found for the 3′ terminus.

Replication models proposed for autonomous parvoviruses included a "rolling hairpin" scheme (Fig. 11-27, p. 402) with dimeric molecules as intermediates. An alternative unidirectional, "semi-discontinuous" model,[19] similar to the presumed mechanism at a papovaviral or eukaryotic chromosomal fork, suffers from the lack of evidence for RNA primers or short "Okazaki" intermediates.[20] Available data also fit with the model proposed for AAV (Fig. S15-2), with modifications to account for the asymmetry in the mechanism involving terminal sequences.

## Replication Proteins and In Vitro Studies

Several viral mRNAs and splice and promoter sites have been identified, but aside from virion structural proteins there is little information about parvoviral gene products. The source and function of the autonomous parvoviral terminal protein (p. 583)[21] are not known, and its function may not mimic that of the adenoviral terminal protein in initiation of new strands (Section S15-4).

With nuclei or lysates prepared from cells infected with autonomous parvoviruses, DNA polymerase $\alpha$ was required when primarily viral strands were synthesized,[22] whereas both DNA polymerases $\alpha$ and $\gamma$ appeared to be functioning when the product was replicative form DNA.[23] The capacity of purified DNA polymerase $\gamma$ to convert viral single strands to double-stranded replicative form DNA[24] is consistent with a role in the synthesis of (i) parental replicative form DNA from an infecting viral single strand, and

18. Rhode, S. L., III and Klaassen, B. (1982) *J. Virol. 41*, 990.
19. Rhode, S. L., III (1978) in *Replication of Mammalian Parvoviruses*, CSHL, p. 279.
20. Tseng, B. Y., Grafstrom, R. H., Revie, D., Oertel, W., and Goulian, M. (1979) *CSHS 43*, 263.
21. Revie, D., Tseng, B. Y., Grafstrom, R. H., and Goulian, M. (1979) *PNAS 76*, 5539.
22. Pritchard, C., Stout, E. R., and Bates, R. C. (1981) *J. Virol. 37*, 352.
23. Kollek, R., Tseng, B. Y., and Goulian, M. (1982) *J. Virol. 41*, 982.
24. Kollek, R., and Goulian, M. (1981) *PNAS 78*, 6206.

terminus of parental strand, and chain extension at nick, has resulted in a "hairpin transfer." IV. Reformation of terminal hairpin loops has generated a 3′-primer terminus. V. Synthesis from the 3′-primer terminus has displaced a unit length of genome. VI. Completion of one round of displacement synthesis has resulted in a displaced single strand (ready to start synthesis of a complementary strand, as in step I), and a duplex molecule that can undergo another round of hairpin transfer and displacement synthesis (beginning at step II). Note that steps IV–VI are equally likely to take place from the right-hand end as from the left. [After Berns, K. I., and Hauswirth, W. W. (1982) in *Organization and Replication of Viral DNA* (A. Kaplan, ed.), CRC Press, Cleveland. (Courtesy of Professor M. Goulian.)]

(ii) complementary strands on single-strand templates produced by displacement synthesis of duplex replicative form DNA (Fig. S15-2).

## S15-4   Adenoviruses

Discovery of an active in vitro system for initiation of replication has advanced the elucidation of a novel mechanism for priming new chain synthesis and the interplay of host and viral replication proteins.

### Replication in Vivo[25]

The terminal 55-kdal protein of the virion is derived from an 80-kdal precursor by cleavage late in the infective cycle. The protein is linked by a phosphodiester bond from serine to the 5'-terminal deoxycytidine residue of each strand. In the virion of a protease-deficient mutant (Ads ts1)[26] grown at a nonpermissive temperature, the terminal protein is 80 kdal. Circular structures formed by cohesion of the two 5'-terminal 55-kdal protein molecules observed in the disrupted virion have also been seen among the replicative forms of adenovirus in infected HeLa cell nuclei.[27]

### Replication in Vitro[28]

Enzyme fractions from adenovirus-infected cells support the initiation and extensive replication of an adenoviral DNA template complexed with terminal protein. The latter is dispensable for template function.[28a] All newly synthesized DNA chains are covalently linked at their 5' termini to the 80-kdal precursor of the virion terminal protein. The sequence of reactions may be reconstructed from the following observations:[29]

   (i)   An active complex of the 80-kdal protein-adenoviral DNA with a 140-kdal DNA polymerase could be isolated from cytosol extracts of infected cells. The associated DNA polymerase in the complex resembles DNA polymerase $\alpha$ in its template preference

25. Challberg, M. D., Desiderio, S. V., and Kelly, T. J., Jr. (1980) *PNAS 77*, 5105; Desiderio, S. V., and Kelly, T. J., Jr. (1981) *JMB 145*, 319; Stillman, B. W., Lewis, J. B., Chow, L. T., Mathews, M. B., and Smart, J. E. (1981) *Cell 23*, 497; Lichy, J. H., Horwitz, M. S., and Hurwitz, J. (1981) *PNAS 78*, 2678.
26. Weber, J. (1976) *J. Virol. 17*, 462.
27. Girard, M., Bouché, J. P., Marty, L., Revet, B., and Berthelot, N. (1977) *Virol. 83*, 34.
28. Challberg, M. D., Desiderio, S. V., and Kelly, T. J., Jr. (1980) *PNAS 77*, 5105; Lichy, J. H., Horwitz, M. S., and Hurwitz, J. (1981) *PNAS 78*, 2678; Challberg, M. D., Ostrove, J. M., and Kelly, T. J., Jr. (1982) *J. Virol. 41*, 265.
28a. Tamanoi, F., and Stillman, B. W. (1982) *PNAS 79*, 2221.
29. Ikeda, J.-E., Enomoto, T., and Hurwitz, J. (1981) *PNAS 78*, 884; Enomoto, T., Lichy, J. H., Ikeda, J.-E., and Hurwitz, J. (1981) *PNAS 78*, 6779.

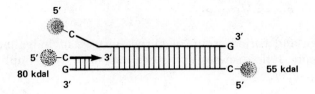

FIGURE S15-3
Scheme for initiation of adenoviral DNA replication. A virus-encoded 80-kdal protein is covalently attached to the 5'-P start of the nascent chain; the 80-kdal polypeptide is processed to 55 kdal before the DNA is packaged. Recent studies [Tamanoi, F., and Stillman, B. W. (1982) *PNAS 79*, 2221] show that in the absence of the 55-kdal protein, the 3'-terminal sequence of a single-stranded template supports initiation.

and sensitivity to N-ethylmaleimide, araCTP, and salt, but differs in its insensitivity to aphidicolin.

(ii)   The 80-kdal protein in the complex is covalently linked to dCMP in an aphidicolin-insensitive reaction[30] with dCTP, ATP, and $Mg^{2+}$ (Fig. S15-3). A nuclear extract from uninfected (or infected) cells increases the rate of reaction tenfold. A 2-kb restriction fragment of adenoviral DNA containing the terminal protein is as effective as the full-length, 30-kb genome.[31]

(iii)   The 80-kdal protein-dCMP complex is extended to a 26-nucleotide-long chain by adding dCTP, dATP, dTTP, and dideoxy-GTP, because the first cytosine residue in the template does not appear until position 26. This chain growth, as well as the initiation reaction, is insensitive to aphidicolin inhibition.

(iv)   Full-length adenoviral DNA is synthesized with the further addition of the virus-encoded 72-kdal DNA-binding protein and a cytosol fraction from uninfected cells that may supply helicase activity. Extensive replication is sensitive to aphidicolin inhibition.

In the least complicated interpretation of these results, the 80-kdal protein starts new DNA chain synthesis by incorporating the 5'-dCMP terminal residue, which in turn provides the primer-terminus for replication. However, until the reactants in this system are more refined, alternative mechanisms cannot be excluded.[32] The unusual nature of the strange DNA polymerase also needs to be clarified.

30. Pincus, S., Robertson, W., and Rekosh, D. (1981) *N. A. Res. 9*, 4919.
31. Horwitz, M. S., and Ariga, H. (1981) *PNAS 78*, 1476.
32. Harpst, J. A., Chow, K., Corden, J. L., and Pearson, G. D. (1981) *J. Supramol. Structure and Cell. Biochem.* Abstract no. 962 of 10th Annual ICN–UCLA Symposium.

## S15-5   Herpes Viruses

The large and varied group of herpes viruses includes many that cause acute, lytic diseases and some that have been implicated in causing cancer.

### Virion and Genome

The icosahedral viral capsid, about 100 nm in diameter, is enveloped in a membranous shell 120 to 150 nm in diameter and is usually found in the perinuclear space, in the cisternae of the endoplasmic reticulum, and in extracellular fluids.

The linear duplex genome of about 150 kb varies in GC content along its length from 32 to 75 percent and encodes about 50 viral polypeptides. The currently preferred structure of the genome contains two covalently linked large and small components (L and S) with predominantly unique sequences ($U_L$ and $U_S$) each bracketed by large, inverted repeats ($ab$ and $b'a'$; $a'c'$ and $ca$):[33]

$$a_{(1 \text{ to } 14)}\, b\, U_L\, b'a' \,\ldots\, a'c'\, U_S\, ca$$

The $a$ sequence ranges from 200 to 500 bp in different viruses and is present as one or many tandem copies. Although the orientation of the L and S components can be inverted relative to one another, the $a$ sequence is oriented similarly at both ends of the genome. The $a$ sequence enables the genome to circularize or concatemerize through exonucleolytic exposure of complementary cohesive termini. It is also the region for the site-specific recombinations that generate the inversions of the L and S components.

Interruptions in the phosphodiester backbone of herpes simplex DNA are now attributed to nicks or gaps rather than intermittent ribose residues.[34]

### Infection and Replication

Whether infection takes an active or latent course depends on the virus and host cell. Very likely, the viral genome encodes information designed to establish latency and thus ensure survival of the virus in nature. Some replicative cycles, such as those of herpes simplex virus (HSV) and also Epstein-Barr virus are rapid and completed in 20 hours; the cytomegalovirus (CMV) cycle is relatively

---

33. Roizman, B. (1979) *Cell 16,* 481; Mocarski, E. S., and Roizman, B. (1981) *PNAS 78,* 7047.
34. Ecker, J. R., and Hyman, R. W. (1981) *Virol. 110,* 213.

slow. The nascent HSV DNA is endless because it is circular or a long concatemer,[35] but the molecular details of its replication have not been established.

_Cell transformation._ The capacity of fragments of the herpes genome to transform cells to an oncogenic phenotype is consonant with the failure to find an entire HSV genome in a transformed cell. Furthermore, the sequences that transform in vitro are different from those that are expressed or found in human tumor tissues. Of great current interest is the use of herpes genes, particularly the thymidine kinase gene, to convert and select cells expressing this acquired gene and others linked to it.[36]

## S15-6 Pox Viruses: Vaccinia

The very large (180 kb) duplex genome of vaccinia contains at each of its continuous (crosslinked) ends (Fig. 15-10) a terminal 10-kb repetition with two sets of direct tandem repeats within it.[37]

The virion is replete with enzymes for the synthesis and processing of RNA. Permeabilized virions afford an attractive in vitro system in which transcripts are guanylylated (capped), methylated, and polyadenylated for translation by a protein synthesizing system in vitro. RNA polymerase has been purified extensively,[38] and a closer look is now possible at the fascinating dependence of transcription on ATP hydrolysis. A coupled action of phosphohydrolase II of the virion may occur here in a manner analogous to involvement of the DNA-dependent ATPases in replication (Section S9-9).[39]

The DNA polymerase encoded by the virus has been purified to homogeneity (Section S6-7). It is notable for its sensitivity to aphidicolin and possession of a $3' \rightarrow 5'$ exonuclease activity. Induced proteins related to replication (p. 595) include a topoisomerase[40] and single-strand binding protein.[41]

Replication in the cytosolic factories, called "virosomes," appears to follow a hairpin-loop mechanism (Fig. 15-10)[42] with clear analogies to those proposed for the parvoviruses (Section S15-3).

35. Jacob, R. J., Morse, L. S., and Roizman, B. (1979) _J. Virol. 29_, 448.
36. Post, L. E., Mackem, S., and Roizman, B. (1981) _Cell 24_, 555.
37. Wittek, R., Cooper, J. A., and Moss, B. (1981) _J. Virol. 39_, 722; Baroudy, B. M., Venkatesan, S., and Moss, B. (1982) _Cell 28_, 315.
38. Spencer, E., Shuman, S., and Hurwitz, J. (1980) _JBC 255_, 5388.
39. Shuman, S., Spencer, E., Furneaux, H., and Hurwitz, J. (1980) _JBC 255_, 5396.
40. Bauer, W. R., Resner, E. C., Kates, J., and Patzke, J. V. (1977) _PNAS 74_, 1841.
41. Nowakowski, M., Kates, J., and Bauer, W. (1978) _Virol. 84_, 260.
42. Moyer, R. W., and Graves, R. L., personal communication.

# S15-7   Retroviruses (RNA Tumor Viruses)[43]

## Genome

Extensive regions of murine leukemia virus genomes and the entire Rous sarcoma virus genome have been sequenced.[44,45] Topographical features include (Fig. S15-4): (i) signals that are essential to the genesis of viral mRNAs; (ii) the boundaries of all viral genes; (iii) a short overlap between the *pol* and *env* genes (with the coding sequences of the two genes in different translational frames); (iv) a previously conjectured but unsubstantiated requirement for splicing in order to express *pol;* and (v) the basis for the first amino acid sequence for a retroviral transforming protein ($pp60^{v-src}$).

Additional tRNAs that prime viral DNA synthesis have been identified for several retroviruses, including mouse mammary tumor virus (tRNA$^{Lys}$) and murine leukemia and reticuloendotheliosis viruses (tRNA$^{Pro}$). Features of the viral genomes determine which species of tRNA is used as primer rather than the host cells in which the viruses replicate.

More than fifteen distinct retrovirus oncogenes have now been identified. All but one of these apparently originated by transduction of normal genes from the genomes of vertebrate species. In most if not all instances, the viral gene and its cellular progenitor are closely related in structure and function.[46] Some oncogenes resemble *src* in their biological effects and biochemical mechanisms, and others appear to be entirely different. The protein kinases encoded by *src* ($pp60^{v-src}$) and by its normal cellular homologue ($pp60^{c-src}$), and perhaps the proteins encoded by several other oncogenes, phosphorylate tyrosine in protein substrates,[47] a reaction of possibly wide regulatory importance in normal and neoplastic cells.

## Proviral State

The integrated DNAs of retroviruses bear striking resemblances to the DNAs of transposable elements of bacteria, yeast, and *Drosophila*:[48] (i) the provirus is flanked by short (4–6 bp) direct repetitions of cellular DNA that are apparently produced from a single copy of

43. *Molecular Biology of RNA Tumor Viruses* (1980) (Stephenson, J. R., ed.) Academic Press, New York; *RNA Tumor Viruses* (1982) (Weiss, R. A., Teich, N. M., Varmus, H. E., and Coffin, J. M., eds.) CSHL; Bishop, J. M. (1982) *Sci. Amer. 246*, No. 3, 80.
44. Van Beveren, C., Rands, E., Chattopadhyay, S. K., Lowy, D. R., and Verma, I. M. (1982) *J. Virol., 41*, 542.
45. Schwartz, D., and Gilbert, W. (1982) *Cell*, in press.
46. Bishop, J. M. (1981) *Cell 23*, 5.
47. Hunter, T. (1980) *Cell 22*, 647; Smart, J. E., Oppermann, H., Czernilofsky, A. P., Purchio, A. F., Erikson, R. L., and Bishop, J. M. (1981) *PNAS 78*, 6013.
48. Temin, H. M. (1980) *Cell 21*, 599.

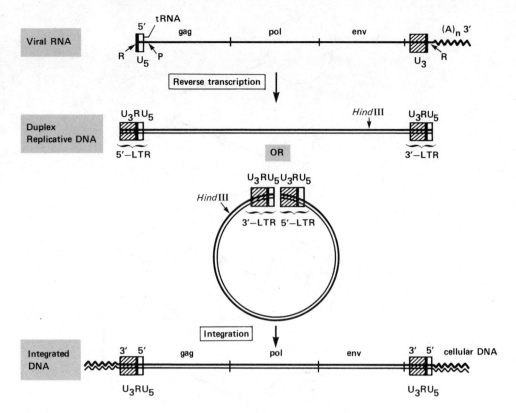

Figure S15-4
Structure of retroviral genomic RNA, duplex forms of replicative DNA, and integrated viral DNA. The 10-kb RNA strand is capped at the 5′ end and terminated with a polyadenylate tail of about 200 residues at the 3′ end. A short terminal repetition or redundancy (R) is found at each end. P is the binding site for the primer tRNA. The unique sequences, $U_5$ and $U_3$, together constitute a long-terminal repeat (LTR) at the termini of the duplex viral DNA. Gene order is *gag* (structural proteins of viral core), *pol* (reverse transcriptase), *env* (glycoprotein(s) of viral envelope), and in some retroviruses, *src* (kinase for cell transformation). All integrated viral DNA and some of the replicative duplex DNA forms are several hundred nucleotides longer than genomic RNA by virtue of having two LTRs. (Adapted from Professor J. M. Bishop and Dr. I. M. Verma.)

the sequence during the act of integration; (ii) the provirus proper displays large direct terminal repeats ("LTR", the $U_3$-R-$U_5$ unit of 270–1330 bp), which are themselves bounded by smaller (3–21 bp) inverted terminal repeats; and (iii) two nucleotides are apparently lost from each end of viral DNA at the time of integration, after which proviruses of diverse viral strains all terminate with characteristic dinucleotides (5′-TpG . . . CpA-3′). Despite these provocative similarities, there is as yet no evidence that proviruses can transpose directly from one position in host cell DNA to another

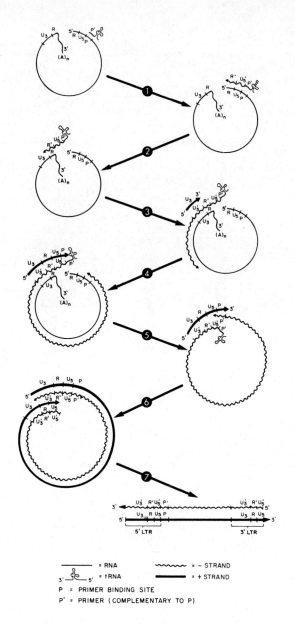

FIGURE S15-5
Scheme for reverse transcription of a viral RNA genome to a linear duplex replicative DNA form.

Step 1: Initiation of minus strand (cDNA). Synthesis of a viral DNA (−) strand (cDNA) initiates at the 3′-OH end of the tRNA primer (designated P′) annealed to P located near the 5′ end of the genomic RNA. Synthesis proceeds about 150 nucleotides until it reaches the 5′ terminus of the genomic RNA.

Step 2: Minus strand jump. The started cDNA (R′U$_5$′) covalently linked to the primer tRNA is released (perhaps by RNase removal of the genomic RNA) and "jumps" to hybridize with the available terminally redundant sequence at the 3′ end of the RNA. Synthesis of cDNA then proceeds toward the 5′ end of the RNA.

without exploiting an RNA intermediate and reverse transcription, nor is there evidence for the attractive possibility that retroviruses function as generalized transducing agents.

## Replication

RNA-directed DNA synthesis remains unique to the life cycle of retroviruses (reverse transcriptase viruses); no similar reaction is known in uninfected cells. Reverse transcriptase is a versatile protein, with as many as three distinct functional domains that mediate DNA synthesis, RNaseH activity, and nucleic acid binding (Section S6-8).

In current schemes for viral DNA synthesis (Fig. S15-5),[49] five domains of the viral genome are crucial: $U_5$, 80–120 nucleotides at the 5' terminus of the genome; $U_3$, 170–1200 nucleotides at the 3' terminus of the genome; R, a direct redundancy of 13–80 nucleotides at both ends of the viral genome; P, a sequence of about 18 nucleotides that base pairs with the 3'-stem region of the tRNA primer; and a short sequence of purines at the 5' boundary of $U_3$. The model accounts for the genesis of LTRs, the replication of both ends of a linear template without sacrifice of genetic content, the initiation and propagation of both strands of viral DNA, and the role of the terminal redundancy (R) in viral DNA synthesis.

The most perplexing aspect of viral DNA synthesis is how the second (+) strand is initiated, propagated, and completed. Whether synthesis is by uninterrupted chain propagation,[50] or as previously proposed, by discontinuous synthesis,[51] is still unclear. The primer for plus strand synthesis, presumed to be viral RNA, has not been identified, but the site of initiation has been located at the precise

---

49. Gilboa, E., Mitra, S. W., Goff, S., and Baltimore, D. (1979) *Cell 18*, 93.
50. Gilboa, E., Mitra, S. W., Goff, S., and Baltimore, D. (1979) *Cell 18*, 93.
51. Kung, H. J., Fung, Y. K., Majors, J. E., Bishop, J. M., and Varmus, H. E. (1981) *J. Virol. 37*, 127.

---

*Steps 3 and 4: Initiation of (+) strand.* Replication of the cDNA occurs over a template that includes $U_3'$, R', and $U_5'$ and also the P' region of the tRNA. The cDNA itself continues to elongate and encompasses the primer tRNA binding site (P of the genomic RNA).

*Steps 5–7: Removal of RNA, plus-strand jump, and completion of linear duplex DNA.* Precisely how and when in these and preceding steps the genomic RNA is removed, presumably by the RNaseH activity of the reverse transcriptase, is still uncertain. The P region on the (+) strand can pair with P' on the cDNA, transcribed from genomic RNA. Synthesis of the cDNA then continues with replication off the (+) strand and the latter elongates by using the cDNA as template. The final duplex DNA has long terminal repeats and is longer by an LTR length than the genomic RNA.

[Based on Gilboa, E., Mitra, S. W., Goff, S., and Baltimore, D. (1979) *Cell 18*, 93. (Courtesy of Dr. I. M. Verma.)]

leftward boundary of $U_3$, where a highly conserved polypurine tract is found in the genomes of widely divergent strains of retroviruses.[52] The tRNA primer for minus strand synthesis has acquired further importance in present models of viral DNA synthesis. Transcription of (+) strand DNA from the 3' stem of the tRNA on the one hand, and of (−) strand DNA from the primer binding site in the genome on the other hand, would create an intermediate form of linear duplex DNA with cohesive ends following removal of RNA by RNaseH. Annealing of the ends would facilitate chain propagation to complete both the (−) and (+) strands, and also generate the characteristic configuration of LTRs (Fig. S15-5).

The final linear products of viral DNA synthesis are duplexes bounded by LTRs (whose composition can be written $5'$-$U_3$-R-$U_5$-$3'$). The exact configuration of the ends of the linear forms—whether "blunt" or "overhanging"—has not been determined. Circular duplexes of viral DNA are formed later, apparently by two mechanisms: (i) blunt-end ligation of the linear forms with no loss of sequence, or (ii) homologous recombination between the LTRs, leaving a single complete LTR in the circle.[53,54] The enzymatic mechanisms that mediate these events have not been identified and presumably are provided by the host cell.

Reverse transcription within purified, permeabilized virions can produce infectious viral DNA, complete with LTRs,[55] and even circular forms.[56] Similar results have not been achieved with purified reverse transcriptase, perhaps for lack of unidentified but essential cofactors, such as nucleic acid binding proteins.


Integration

The form of viral DNA integrated into the host chromosome remains unknown. Current views of mechanism are based on two sources:[57] arguments by analogy with the models for transposition of DNA in bacteria, and "cointegrate" structures that arise when unintegrated molecules of viral DNA recombine in a manner that is alleged to imitate integration. Two features of integration are clear:[58] (i) the

52. Swanstrom, R., DeLorbe, W. J., Bishop, J. M., and Varmus, H. E. (1981) PNAS 78, 124; Majors, J. E., and Varmus, H. E. (1981) Nat. 289, 253.
53. Dhar, R., McClements, W. L., Enquist, L. W., and Vande Woude, G. F. (1980) PNAS 77, 3937.
54. Swanstrom, R., DeLorbe, W. J., Bishop, J. M., and Varmus, H. E. (1981) PNAS 78, 124.
55. Gilboa, E., Goff, S., Shields, A., Yoshimura, F., Mitra, S., and Baltimore, D. (1979) Cell 16, 863.
56. Clayman, C. H., Mosharrafa, E. T., Anderson, D. L., and Faras, A. J. (1979) Science 206, 582.
57. Shoemaker, C., Goff, S., Gilboa, E., Paskind, M., Mitra, S. W., and Baltimore, D. (1980) PNAS 77, 3932.
58. Temin, H. M. (1980) Cell 21, 599; Majors, J. E., and Varmus, H. E. (1981) Nat. 289, 253; Shimotohno, K., and Temin, H. M. (1980) PNAS 77, 7357; Hughes, S. H., Mutschler, A., Bishop, J. M., and Varmus, H. E. (1981) PNAS 78, 4299.

participating site on viral DNA is highly specific, whereas a large number of apparently different sites are used in cellular DNA; and (ii) there is no apparent requirement for homology between the interacting sites on viral and cellular DNA. Neither viral DNA synthesis nor integration can occur in the absence of cellular DNA synthesis, but the precise contribution of the host cell to these two crucial events in retrovirus replication remains unknown.[59]

The provirus appears to be a self-contained transcriptional unit,[60] with a characteristic site for initiation of RNA synthesis, canonical signals that direct the splicing of RNA during the genesis of viral mRNAs, a signal for polyadenylylation of the completed mRNAs, and—in the case of mouse mammary tumor virus—a domain that mediates the stimulation of viral RNA synthesis by steroid hormones.[61] Cellular DNA adjacent to the provirus may nevertheless modulate or even completely repress the synthesis of viral RNA. As a consequence, the site of integration on the cellular genome can be an important determinant of viral gene expression.

## S15-8   Mitochondria, Chloroplasts, and Plasmids

### Mitochondria

The several thousand copies of closed, circular, mitochondrial DNA in a mammalian cell are autonomously replicated and transcribed within the organelle. The complete protein-synthesizing system depends on 22 tRNAs, 12S and 16S ribosomal RNAs, cytochrome b, three of the cytochrome oxidase subunits encoded in the mitochondrial genome, and many imported nuclear gene products. With sequencing of the genome now completed, the origins of mitochondrial DNA replication are sharply defined (see below), but still virtually nothing is known about the enzymology of any stage of replication.

_Genome_.[62]   Early revelations from sequencing of mitochondrial genomes were a genetic code and decoding mechanism[63] distinctly different from those established for nuclear and prokaryotic ge-

59. Humphries, E. H., Glover, C., and Reichmann, M. E. (1981) _PNAS 78_, 2601.
60. Hackett, P. B., Varmus, H. E., and Bishop, J. M. (1981) _Virol. 112_, 714.
61. Ringold, G. M. (1979) _BBA 560_, 487.
62. _Mitochondrial Genes_ (1982) (Slonimski, P., and Attardi, G., eds.) CSHL.
63. Sanger, F. (1981) _Science 214_, 1205; Macino, G., and Tzagoloff, A. (1979) _PNAS 76_, 131.

nomes. The total sequences of the mouse (16,295 bp)[64] and human (16,569 bp)[65] genomes show also their extensive homology and great economy of organization. Genes are separated by few or no non-coding bases between them. The only two significant portions of the mouse genome that do not encode a functional RNA species are the 879-bp displacement-loop region containing the template origin for H-strand replication and the 32-bp template origin for L-strand replication. In many instances, termination codons are not included in the DNA, but are defined instead by polyadenylylation of mRNA posttranscriptionally.

Alkali-sensitive sites found in ten or more places throughout mitochondrial genomes have been more precisely located by the positions of substituted ribonucleotides.[66] Their retention at the two origins of replication and at other preferred sites confirms suggestions that they are vestiges of RNA priming events in DNA replication. Survival of integrated ribonucleotides is evidence for the lack in mitochondria of DNA repair systems that remove such remnants from nuclear and prokaryotic chromosomes.

The long (43 kb) linear mitochondrial DNA of *Paramecium*, in which crosslinking of the duplex strands at one end starts replication, will entail novel mechanisms for initiation and termination.[67]

*Replication*.[68]   The displacement-loop region in mouse and human mitochondrial DNA containing the origin of H-strand synthesis is the most divergent in the genome, with striking lack of homology at both the 5′ and 3′ ends of the region;[69] on the other hand, the sequence of L-strand origins is highly conserved between species and contains an unambiguous hairpin structure. A short repeated sequence in the L-strand template at the 3′ end of displacement-loop strands may serve in arresting their synthesis.[70] Replicative intermediates (D-loops) are stabilized by a 16-kdal polypeptide and thus protected from being lost by branch migration of the displaced strand.[71]

The mitochondrial system, by affording a well-defined template, an isolatable organelle, and a characterized replicase (e.g., DNA

64. Bibb, M. J., Van Etten, R. A., Wright, C. T., Walberg, M. W., and Clayton, D. A. (1981) *Cell 26*, 167.
65. Anderson, S., Bankier, A. T., Barrell, B. G., de Bruijn, M. H. L., Coulson, A. R., Drouin, J., Eperon, I. C., Nierlich, D. P., Roe, B. A., Sanger, F., Schreier, P. H., Smith, A. J. H., Staden, R., and Young, I. G. (1981) *Nat. 290*, 457.
66. Brennicke, A., and Clayton, D. A. (1981) *JBC 256*, 10613.
67. Pritchard, A. E., and Cummings, D. J. (1981) *PNAS 78*, 7341.
68. Clayton, D. A. (1982) *Cell*, in press.
69. Bibb, M. J., Van Etten, R. A., Wright, C. T., Walberg, M. W., and Clayton, D. A. (1981) *Cell 26*, 167; Tapper, D. P., and Clayton, D. A. (1981) *JBC 256*, 5109; Walberg, M. W., and Clayton, D. A. (1981) *N. A. Res. 9*, 5411.
70. Doda, J. N., Wright, C. T., and Clayton, D. A. (1981) *PNAS 78*, 6116.
71. Van Tuyle, G. C., and Pavco, P. A. (1981) *JBC 256*, 12772.

polymerase $\gamma$), still beckons enzymologists to define the detailed molecular mechanisms and their regulation.

## Plasmids

Focusing on autonomously replicating yeast plasmids as guides to mechanisms of nuclear DNA replication (Section S13-6) is justified by their genetic stability, physiological responses to nuclear regulatory controls, and results from in vitro systems in which replication is initiated at a specific origin used in vivo and proceeds bidirectionally from that point. The soluble enzyme system provides a basis for assaying products of cell-development-cycle genes as well as novel replication factors and accessory proteins for DNA polymerase.

*2-$\mu$m yeast plasmid.*[72]   This 6318-bp extrachromosomal circle, present in about fifty copies per cell, is under the same strict cell cycle controls as those that govern replication of chromosomal DNA. A soluble extract of yeast replicates both the plasmid and the chimeric plasmids that contain it.[73] Initiation is selective within one of the 599-bp inverted repeats to form a typical "eye" form. Replication is bidirectional, goes to completion, and provides an assay for the product of *cdc8,* one of the cell-development-cycle genes.

*Recombinant ars plasmid.*   A minute plasmid (1453 bp)[74] fashioned from a yeast tryptophan gene (*trp1*) and *ars1* (an autonomously replicating sequence)[75] is stably maintained at 100–200 copies per cell through mitotic and meiotic cycles. Replication depends, as does that of nuclear DNA (but not mitochondrial DNA), on *cdc* genes (28, 4, and 7 function prior to S phase; 8 and 21 operate during S phase) and the inhibitory effect of the yeast pheromone $\alpha$-factor. The typical nucleosomal organization of the plasmid and its physiological behavior make it a desirable object for intensive genetic and enzymological analysis.

A soluble enzyme system from yeast that replicates recombinant plasmids with an *ars* starts at that sequence and proceeds bidirectionally.[76] As in studies with the 2-$\mu$m plasmid,[73] synthesis is defective with extracts prepared from *cdc8* mutant cells and can be restored by complementation with a partially purified cdc8 protein.

72. Broach, J. R. (1982) *Cell 28,* 203.
73. Kojo, H., Greenberg, B. D., and Sugino, A. (1981) *PNAS 78,* 7261.
74. Zakian, V. A., and Scott, J. F. (1982) *Mol. Cell. Biol. 2,* 221.
75. Stinchcomb, D. T., Thomas, M., Kelly, J., Selker, E., and Davis, R. W. (1980) *PNAS 77,* 4559; Stinchcomb, D. T., Mann, C., and Davis, R. W., *JMB,* in press.
76. Campbell, J. L., personal communication.

# S15-9   Hepadna Viruses:[77] Hepatitis B (HBV)

Hepadna viruses are distributed worldwide in humans with 20 percent of the population in some areas actively infected. The occurrence of the integrated virus in hepatoma cell cultures[78] and close association with chronic hepatitis and hepatocellular carcinomas[79] focus attention on understanding the replicative cycle of HBV. Finding viruses closely related to HBV in woodchucks, ground squirrels,[80] and ducks is the basis for assembling them in a new hepadna group. A novel organization of a tiny genome and a virus-associated DNA polymerase pose an interesting problem in replication.

## Virion and Genome

The HBV virion, also known as the Dane particle, is a 42-nm spherical particle with a lipid envelope bearing the hepatitis B surface antigen ($HB_sAg$); a 28-nm nucleocapsid core contains the core antigen ($HB_cAg$), the viral DNA, a DNA polymerase activity, and probably the $B_e$ antigen in a cryptic form. These features and polypeptides characterize the hepadna virus group. In persistent human infections, high concentrations of incomplete viral forms, as well as intact particles, are found in the blood. These include small $HB_sAg$ particles 16 to 25 nm wide and long $HB_sAg$ filaments (22 nm wide by 500 nm long).

The entire sequence of the viral DNA, a partially duplex circle, is known.[81] The longer strand of 3182 nucleotides, matched for 1700 to 2800 bases by the complementary strand in different molecules, is nicked at a point 300 nucleotides from the 5′ end of the shorter strand. The 5′ terminus of the long strand has a covalently attached protein;[82] the 5′ end of the short strand has no protein attached but is not phosphorylatable.

## Replication and Integration

The DNA in isolated virions, in the presence of deoxynucleoside triphosphates, is repaired by the endogenous DNA polymerase to a

77. Robinson, W. S., Marion, P., Feitelson, M., and Siddiqui, A. (1982) in Proceedings of the 1981 Symposium on Viral Hepatitis (Altic, H., Maynard, J., and Szmuness, W., eds) Franklin Institute Press, Philadelphia, in press.
78. Marion, P., Salazar, F. H., Alexander, J. J., and Robinson, W. S. (1980) J. Virol. 33, 795.
79. Bréchot, C., Hadchouel, M., Scotto, J., Fonck, M., Potet, F., Vyas, G. N., and Tiollais, P. (1981) PNAS 78, 3906.
80. Marion, P. L., Oshiro, L. S., Regnery, D. C., Scullard, G. H., and Robinson, W. S. (1980) PNAS 77, 2941.
81. Galibert, F., Mandart, E., Fitoussi, F., Tiollais, P., and Charnay, P. (1979) Nat. 281, 646; Galibert, F., Chen, T. N., and Mandart, E. (1982) J. Virol. 41, 51.
82. Gerlich, W. H., and Robinson, W. S. (1980) Cell 21, 801.

nearly complete duplex circle of 3200 bp. Controlled heating converts the duplex to a linear form, indicating that a nick or gap remains in the repaired short strand and that a nearby nick or gap is also present in the long strand. The DNA polymerase has not been separated from the virion and so its origin, structure, and properties are unknown. The protein covalently bound at the 5' end of the constant length strand may play a replicative role, but no defined size or function has yet been established. In subviral particles of infected duck liver, synthesis of the long strand appears to utilize an RNA template, implicating an initial transcription of the infecting HBV DNA, followed by reverse transcription of the RNA to produce progeny.[83] The possibility that the template for transcription is integrated viral DNA also exists. Similarities with retroviruses (Section S15-7) are striking: reverse transcription, integration, and terminal repeats in linearized HBV DNA resembling retroviral LTR sequences.

83. Mason, W. S., Aldrich, C., Summers, J., and Taylor, J. M. (1982) *PNAS 79*, in press; Summers, J., and Mason, W. S. (1982) *Cell*, in press.

# Repair, Recombination, Transformation, Restriction, and Modification

## Abstract

The ways in which DNA is modified, repaired, and recombined, even within a single cell, continue to multiply and impress the observer with the dynamic character of the once staid DNA molecule. With improved methods for detecting DNA lesions, enzymatic machinery for dealkylation and specific N-glycosylases have been discovered. In the so-called SOS state, with its many inducible functions, the mechanism of "error-prone" (translesion) repair is still unknown. Although some human diseases are due to a defect in DNA repair, correlations of diminished repair with aging remain to be established.

Intensive studies of recA protein are making the once mysterious generalized recombination nearly as plain in molecular detail as a well-worn metabolic pathway. The capacity of this 40-kdal protein to orchestrate the union of vast stretches of DNA and in another guise to sense unpaired DNA strands as a signal for proteolytic cleavage of a selected few regulatory proteins is astonishing.

One of the most active subjects in the biology of DNA is "jumping genes," the site-specific recombinations that are now known to be common and widespread in nature. Transpositions, first appreciated in corn, account for phase variation in *Salmonella,* are a major source of the spontaneous mutations in *Drosophila,* and are common events in yeast. Transpositions are also the basis for oncogenes and integration of tumor retroviruses in animal cells, and judging from the abundance of plasmids, transposons, and insertion elements, provide a popular pastime for DNA in bacteria. Based on the patterns of nucleotide sequences surrounding transpositions, a common theme appears to connect a variety of mechanisms.

With a better understanding of natural systems of bacterial transformation and improved techniques for introducing and assimilating foreign DNA into virtually any cell, the redesign of species is keeping pace with the rearrangement of genes by chemical engineering. Endogenous methylations of the resident DNA in some instances prevent the intrusion of foreign DNA while also assisting with the repair of mismatched bases introduced in replication. But the massive and fluctuating levels of methylation of mammalian DNA seem engaged in other purposes with indications of regulating the expression of genes for major metabolic directions and cellular differentiation.

## S16-1   DNA Damage and Mutations

### Detection of DNA Damage

Electrophilic reactants that bind covalently to DNA bases are generally mutagenic and carcinogenic.[1] Such chemical damage to DNA may be an important cause of human cancer and other diseases. A sensitive and readily applicable test[2] has been developed to detect the presence in DNA of bases altered by a large variety of chemical agents (e.g., N-methyl-N-nitrosourea, dimethyl sulfate, formaldehyde, $\beta$-propiolactone, propylene oxide, streptozotocin, nitrogen mustard, 1-3-bis(2-chloroethyl)-1-nitrosourea). DNA digested enzymatically to 3'-deoxynucleotides is phosphorylated with ($\gamma$-$^{32}$P)ATP and phage T4 polynucleotide kinase. High-resolution chromato-

1. Ames, B. N. (1979) *Science 204,* 587; Hollstein, M., McCann, J., Angelosanto, F. A., and Nichols, W. M. (1979) *Mutat. Res. 65,* 133; International Agency for Research on Cancer (1980) "Long-Term and Short-Term Screening Assays for Carcinogens. A Critical Appraisal." Lyon, France. Suppl. 2.
2. Randerath, K., Reddy, M. V., and Gupta, R. C. (1981) *PNAS 78,* 6126.

DNA nucleotides as an extra and distinctive $^{32}$P-nucleoside diphosphate. It is therefore simple to screen many chemicals for their capacity to bind and damage DNA without having each of them in a highly radioactive form to begin with.

## Probing Sites of DNA Damage in Vivo

Defined sequences of DNA have been used as probes to analyze the spectrum and distribution of damage produced by various deleterious agents. The highly reiterated $\alpha$ DNA sequence that comprises 1 percent of human DNA has been used as an endogenous probe to demonstrate that the distribution of UV light-induced cyclobutyl dimers in the intact chromatin in cells is similar to that in irradiated naked DNA.[3] Furthermore, with this approach a new type of photodamage was observed at cytidine positions to the 3' side of another pyrimidine.

## Mutations from Depurination

Some potent carcinogens (e.g., aflatoxin $B_1$, 2-acetylaminofluorene, benzo[a]pyrene) form bulky adducts to DNA purines and block DNA replication. Also, the N-glycosylic bond of a purine base altered at the N-3 or N-7 position is destabilized, leading to the release of the base. The resulting apurinic sites (AP sites), as well as those produced by heat or acid, are mutagenic during replication in vitro,[4] especially when expressed in an SOS-induced, error-prone state (Sections S16-2, S16-4).[5] These results, and the correlation between the activity of a carcinogen in DNA depurination and missense mutagenicity,[6] increase the possible significance of depurination as a basis for carcinogenesis associated with mutagenesis.

## S16-2   Repair Mechanisms

Repair mechanisms can be subclassified into those that reverse the damage and those that replace the entire damaged unit. Thus, photoreactivation catalyzes direct conversion of a dimer to its constituent pyrimidines, while excision-repair effectively replaces the dimer with the appropriate new pyrimidines.

3. Lippke, J. A., Gordon, L. K., Brash, D. E., and Haseltine, W. A. (1981) *PNAS 78*, 3388.
4. Kunkel, T. A., Shearman, C. W., and Loeb, L. A. (1981) *Nat. 291*, 349.
5. Schaaper, R. M., and Loeb, L. A. (1981) *PNAS 78*, 1773.
6. Drinkwater, N. R., Miller, E. C., and Miller, J. A. (1980) *B. 19*, 5087.

Enzymatic Photoreactivation

Photoreactivating enzymes[7] catalyze the light-dependent repair of cyclobutyl pyrimidine dimers in DNA. A stable enzyme-substrate complex that absorbs light in the 300–600 nm range leads to the conversion of dimers back to the monomers. These photolyases are ubiquitous in their distribution and have been purified from *E. coli,* yeast, *Streptomyces,* algae, and mammals, including humans. In some systems, such as yeast, there are two distinct enzymes. The sizes, action spectra, and other features vary; apparently homogeneous 35-kdal protein from *E. coli* contains a variable amount of carbohydrate and a 10–15 nucleotide stretch of RNA.

## Repair Directly by Dealkylation[8]

Another example of direct reversal is the methyltransferase reaction in *E. coli* in which an acceptor protein receives a methyl group from the procarcinogenic $O^6$-methylguanine (or an ethyl group from $O^6$-ethylguanine), thereby becoming alkylated at one of its own cysteine residues.[9] Unlike a typical enzyme, the acceptor protein is inactivated by the alkylation event. Cultures of *E. coli* conditioned by exposure to low levels of an alkylating agent such as N-methyl-N'-N-nitrosoguanidine become resistant to challenge by a high level of such agents through a 30-fold increase in intracellular levels of this protein. The inducible activity does not appear to be under control of the *recA lexA* regulatory circuit, however.

## Removal of Altered Bases by N-glycosylases[10]

Excision of an altered base by cleavage of the N-glycosylic bond is carried out by a number of glycosylases, each highly specific for a particular type of damage. A number of new glycosylases, including some that remove N-3 and N-7 methyl and ethyl purines, have been discovered.[11] They augment the now well-known glycosylase-incision endonucleases (the T4-endoV and *M. luteus* UV-endonucleases for pyrimidine dimers, and *E. coli* endonuclease III for dihydrothymine) (Section S10-8). Creation of an AP site by N-glyco-

7. Sutherland, B. M. (1981) *Enzymes 14,* 481.
8. Lindahl, T. (1982) *Prog. N. A. Res.,* in press.
9. Sedgwick, B., and Lindahl, T. (1982) *JMB 154,* 169; Cairns, J., Robins, P., Sedgwick, B., Talmud, P. (1981) *Prog. N. A. Res. 26,* 237; Mehta, J. R., Ludlum, D. B., Renard, A., and Verly, W. G. (1981) *PNAS 78,* 6766.
10. Duncan, B. K. (1981) *Enzymes 14,* 565.
11. Lindahl, T. (1982) *Prog. N. A. Res.,* in press; Laval, J., Pierre, J., and Laval, F. (1981) *PNAS 78,* 852; Singer, B., and Brent, T. P., ibid., 856; Margison, G. P., and Pegg, A. E., ibid., 861; Male, R., Nes, I. F., and Kleppe, K. (1981) *EJB 121,* 243.

sylase action is followed by incision of the backbone to leave the AP site on the 3′ end at the nick (Fig. S10-1).

## Incision Preceding Repair[12]

The major *E. coli* system, defined by the *uvrA,B,C* genes (Section S10-8), breaks the phosphodiester backbone on the 5′ side of the pyrimidine dimer as the initial step (Fig. S10-2). With the homogeneous uvrA, B, and C proteins in hand, the way in which this reaction is carried out and how that system may be regulated by *recA* and *lexA* should soon be known.

## Replacement of Excised Regions[13]

The short-patch synthesis that completes most excision-repair events in *E. coli* is largely due to the polymerase and 5′→3′ exonuclease activities of DNA polymerase I (p. 615). Defective repair in plasmolyzed cells of *polA* mutants can be corrected by uptake of the purified enzyme,[14] just as treatment of permeabilized *uvrA* mutants with T4 endoV restores their resistance to UV irradiation.

The long-patch excision-repair pathway in *E. coli* is clearly under *recA lexA* control. The long patches evidently occur at a small, defined number of sites such as those near replication forks or in regions of active transcription where the DNA structure may be opened up. The conditions under which long patches are seen correlate with enhanced UV resistance of the cells or of infecting UV-irradiated, duplex DNA viruses. This condition is nonmutagenic and is much more prominent than the minor proportion of cellular resistance that could possibly be due to translesion (SOS, error-prone) synthesis (see below). Long-patch excision-repair may be responsible for the striking enhancement in UV resistance seen in SOS-induced cells.

Knowledge about the mechanism of direct insertion of a purine basé into an AP site by insertases (p. 616)[15] has not been advanced nor have mutants or other evidence been supplied to evaluate their cellular role.

In mammalian cells, there is no evidence for an inducible long-patch excision-repair mode analogous to that in bacterial cells. It is

12. Friedberg, E. C., Bonura, T., Radany, E. H., and Love, J. D. (1981) *Enzymes* 14, 251.
13. Hanawalt, P. C., Cooper, P. K., and Smith, C. A. (1981) *Prog. N. A. Res. 26*, 181; Cooper, P. K. (1981) in *Chromosome Damage and Repair* (Seeberg, E., and Kleppe, K., eds.), Plenum Press, New York, p. 139; (1982) *MGG*, in press.
14. Shimizu, K., Yamashita, K., and Sekiguchi, M. (1981) *BBRC 101*, 15.
15. Livneh, Z., and Sperling, J. (1981) *Enzymes* 14, 549.

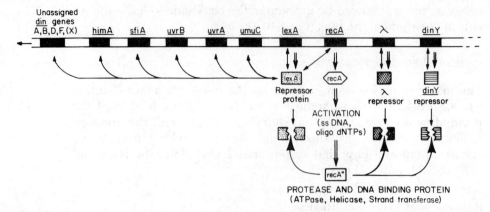

FIGURE S16-1
Regulatory scheme of *recA-lexA* actions in the SOS response in *Salmonella*. The sequence of genes does not represent their order in the chromosome. *din,* damage-inducible genes, are controlled by *lexA* for *dinX* and by a unique, unidentified repressor for *dinY*. (Courtesy of Professor P. Hanawalt.)

likely that bypass pathways (e.g., translesion, error-prone synthesis)[16] are more important in these eukaryotic systems because of the much higher density of growing forks in replicating regions of DNA. There is a greater probability that a lesion will be encountered by a replication fork and fixed in daughter duplexes before it can be repaired by a mechanism acting on the parental DNA.

Further analysis of repair patch size distributions in mammalian cells indicates that this parameter is *not* strongly influenced by the nature of the lesion. No significant variation in patch size is seen following a variety of DNA-damaging treatments when analyzed by the density labeling method. Earlier observations of such variations may have resulted from uncertainties and artifacts in application of the bromouracil-photolysis methodology.

SOS, Error-Prone, Translesion Replication[17]

Among the many reactions induced by DNA damage that establish the SOS state (Fig. S16-1),[18] some lead to an increased number of mutations. These mutations have been attributed to replication

16. Sarasin, A. R., and Hanawalt, P. C. (1980) *JMB 138,* 299.
17. Hall, J. D., and Mount, D. W. (1981) *Prog. N. A. Res. 25,* 53; Howard-Flanders, P. (1981) *Sci. Amer. 245,* No. 5, 72.
18. Little, J. W., Mount, D. W., and Yanisch-Perron, C. R. (1981) *PNAS 78,* 4199; Brent, R., and Ptashne, M. (1981) *PNAS 78,* 4204; Kenyon, C. J., and Walker, G. C. (1980) *PNAS 77,* 2819; (1981) *Nat. 289,* 808; Fogliano, M., and Schendel, P. F. (1981) *Nat. 289,* 196.

across the noninformational site (e.g., AP site) in the template with resultant misincorporations. In the normal, uninduced state, DNA polymerases, as judged by in vitro studies, are arrested at the sites of pyrimidine dimers in the template.[19] Direct evidence has yet to be furnished for translesion synthesis at these sites in the SOS state, although replication of DNA with AP sites is enhanced in spheroplasts from SOS cells as is the frequency of reversions at a specific site.[20]

### Poly(ADP-Ribose) and Repair

DNases and DNA repair have been implicated in stimulating poly-(ADP-ribose) synthesis. The best activation is found with the generation of flush duplex termini containing 5'-hydroxyl groups.[21] Potent inhibition of poly(ADP-ribose) synthetase by 3-aminobenzamide interferes with overall excision-repair without affecting incision or DNA synthesis,[22] suggestive of poly(ADP-ribose) action at an even later stage in repair.

## S16-3   Repair Defects in Disease and Aging

### Disease

Bloom syndrome is a disorder characterized by dwarfism, immune system defects, and an increased incidence of early cancer, chromosomal breaks, and sister-chromatid exchanges. Although DNA repair capacities seem intact, spontaneous mutation rates are elevated fivefold to tenfold in fibroblasts from Bloom patients.[23] These results suggest that Bloom syndrome may be a mutator mutation, a phenomenon previously unknown in humans. Decreased fidelity in replication could account for the developmental abnormalities and predisposition to cancer.

### Aging

The striking correlation of DNA repair capacity with lifespan observed earlier (p. 624) is now controversial as is the general hypothe-

19. Moore, P., and Strauss, B. S. (1979) *Nat. 278*, 664; Moore, P. D., Bose, K. K., Rabkin, S. D., and Strauss, B. S. (1981) *PNAS 78*, 110; Moore, P. D. (1982) *Prog. N. A. Res.*, in press.
20. Schaaper, R. M., and Loeb, L. A. (1981) *PNAS 78*, 1773.
21. Benjamin, R. C., and Gill, D. M. (1980) *JBC 255*, 10493, 10502.
22. Durkacz, B. W., Irwin, J., and Shall, S. (1981) *BBRC 101*, 1433.
23. Warren, S. T., Schultz, R. A., Chang, C.-C., Wade, M. H., and Trosko, J. E. (1981) *PNAS 78*, 3133.

sis that accumulation of errors in proteins, DNA polymerases in particular, are responsible for the phenomena of aging.

In contrast to the correspondence of unscheduled DNA synthesis in fibroblasts from seven species with a 20-fold range in longevity (p. 624), no difference was detected in excision-repair capacity in three cold-blooded vertebrates with lifespans ranging from 3 to over 118 years,[24] nor were significant correlations seen among 34 species in eleven orders of mammals.[25] Excision-repair capacity also does not significantly differ in UV-irradiated human epidermal keratinocytes between newborns and donors up to 88 years of age.[26] On the other hand, the error frequencies of DNA polymerases $\alpha$ and $\gamma$ from senescent human fibroblasts were significantly greater than were those from "young" cells,[27] confirming some previous reports (p. 624).

An interesting hypothesis holds that damage by free-radical reactions is the main cause of aging and age-related disorders,[28] and their prevention by antioxidants is suggested to be of prime importance. If DNA damage proved to be one of the major lesions caused by free radicals, then these effects on DNA and their repair would deserve far more attention.

## S16-4   Recombination Mechanisms

Advances in understanding genetic recombination at a molecular level, limited thus far to prokaryotic systems, have been impressive both for general (homologous) recombination mediated by recA protein and for several varieties of site-specific recombination.

### RecA Protein in Homologous Recombination[29-32]

The extraordinary versatility of modestly sized recA protein was at first hard to believe but now can no longer be doubted. The several recombinase activities as well as the highly specific protease actions

24. Woodhead, A. D., Setlow, R. B., and Grist, E. (1980) *Exp. Geront. 15*, 301.
25. Kato, H., Harada, M., Tsuchiya, K., and Moriwaki, K. (1980) *Japan. J. Genetics 55*, 99.
26. Liu, S., and Hanawalt, P. C., personal communication.
27. Murray, V., and Holliday, R. (1981) *JMB 146*, 55; Krauss, S. W., and Linn, S. (1982) *B. 21*, 1002.
28. Harman, D. (1981) *PNAS 78*, 7124.
29. Weinstock, G. M., McEntee, K., and Lehman, I. R. (1981) *JBC 256*, 8829; 8845; 8850; 8856; McEntee, K., Weinstock, G. M., and Lehman, I. R., ibid., 8835.
30. Shibata, T., Cunningham, R. P., and Radding, C. M. (1981) *JBC 256*, 7557; Shibata, T., DasGupta, C., Cunningham, R. P., Williams, J. G. K., Osber, L., and Radding, C. M., ibid., 7565; Williams, J. G. K., Shibata, T., and Radding, C. M., ibid., 7573; Radding, C. M. (1981) *Cell 25*, 3.
31. McEntee, K., and Weinstock, G. M. (1981) *Enzymes 14*, 445.
32. Dressler, D., and Potter, H. (1982) *ARB 51*, in press.

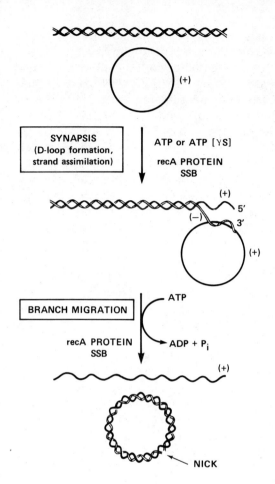

FIGURE S16-2
Stages in homologous recombination catalyzed by recA protein. In this model system, a φX174 viral strand is assimilated into a linear duplex φX174 replicative form by displacement of its viral strand. Although synapsis can occur at either end, the unique polarity of branch migration dictates that displacement is effective from only one end of the duplex. (Courtesy of Professor I. R. Lehman and Professor C. M. Radding.)

reside in the same crystalline 37.8-kdal protein;[33] a single mutation in the structural gene affects both recombinase and protease functions in the isolated protein.[34]

RecA protein pairs a single-stranded or partially single-stranded molecule with an intact duplex molecule of DNA in two experimentally distinguishable steps (Fig. S16-2): *synapsis*, which puts molecules in homologous register by a side-by-side interaction of a single strand with duplex DNA to create a D-loop, and *strand transfer*, which requires a free end and produces long heteroduplex joints by a concerted and polar branch migration. In the synaptic formation

33. Horii, T., Ogawa, T., and Ogawa, H. (1980) *PNAS* 77, 313; Sancar, A., Stachelek, C., Konigsberg, W., and Rupp, W. D. (1980) *PNAS* 77, 2611.
34. Hickson, I. D., Gordon, R. L., Tomkinson, A. E., and Emmerson, P. T. (1981) *MGG* 184, 68.

of D-loops from single-stranded fragments and circular duplex DNA, single strands play a key role in initiating the pairing reaction. Indeed, single-stranded DNA as a cofactor governs all of the activities of recA protein: ATPase, binding of duplex DNA, unwinding of duplex DNA, and protease. The effect of single-stranded DNA in stimulating the binding and partial unwinding of duplex DNA by recA protein occurs whether the single-stranded DNA is homologous or not, which is consistent with a mechanism designed to search for homology. The amount of single-stranded DNA determines the requirement for purified recA protein in the pairing reaction on a stoichiometric basis and the binding to single-stranded DNA appears to be cooperative. Single-strand binding protein has a profound effect in sparing recA protein and may also interact with it.

ATP plays a key role in binding recA protein to single-stranded DNA, but its hydrolysis is essential only in the repeated cycles of detachment and binding of the protein in heteroduplex formation. The kinetics of the pairing reaction resemble classical Michaelis–Menten kinetics. These kinetics distinguish the pairing reaction promoted by recA protein from the pairing of complementary single strands by helix destabilizing proteins.

RecA protein, which lacks detectable topoisomerase activity, will make stable joint molecules from numerous structural variants of DNA if one molecule is at least partially single-stranded, and if a free end exists somewhere in one of the two molecules. Several characterized products of the reaction are illustrated in Figure S16-3. From duplex DNA and fragments of single-stranded DNA, recA protein forms a joint that is a D-loop (Fig. S16-3a); a circular single strand and linear duplex DNA yield joint molecules in which the single strand has displaced one strand of the duplex molecule, creating heteroduplex regions that are hundreds to thousands of base pairs long (Fig. S16-3b); and a linear molecule with single-stranded ends and a duplex molecule that share the same sequence end to end form a Holliday structure in which strands are reciprocally exchanged (Fig. S16-3c).[35]

RecA protein forms less stable joint molecules from circular single-stranded DNA and closed circular duplex DNA, a situation in which a classical heteroduplex joint molecule cannot form for lack of a free end.[36] However, E. coli topoisomerase I, acting with recA protein, converts such an intermediate into a stable catenane in which the circular single strand is topologically linked to its complement in the duplex molecule.[37] These experiments distinguish the

35. DasGupta, C., Wu, A. M., Kahn, R., Cunningham, R. P., and Radding, C. M. (1981) Cell 25, 507.
36. DasGupta, C., Shibata, T., Cunningham, R. P., and Radding, C. M. (1980) Cell 22, 437.
37. Cunningham, R. P., Wu, A. M., Shibata, T., DasGupta, C., and Radding, C. M. (1981) Cell 24, 213.

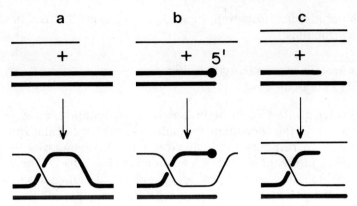

FIGURE S16-3
Alternate mechanisms for homologous recombination catalyzed by
recA protein. See text for details. (Courtesy of Professor C. M.
Radding.)

recognition of sequence homology in synapsis from the actual formation of a classical heteroduplex joint in which the two strands are
helically interwound.

The pairing of circular single-stranded DNA with linear duplex
DNA has provided an ideal system for studying strand transfer and
heteroduplex formation,[38] and the results illustrate the general state-
ments made up to this point. The circular (+) strand of phage DNA
invades and displaces the 5′ end of the (+) strand from linear duplex
DNA by a mechanism that is both concerted and polar. Displacement
of the very end of the (+) strand occurs only when homologous (+)
strands are present, and displacement starts only from the 5′ end of
the (+) strand (Fig. S16-3b; Fig. S16-2); starts from the other end are
abortive. Further growth of these heteroduplex joints, which was
studied in the presence of the single-strand binding protein of E. coli,
requires the continued presence of active recA protein and homology
of the pairing strands; a pyrimidine dimer slows the rate 50-fold.[39]

In vitro, recA protein acts on broken DNA, which is consistent both
with its role in vivo and with the stimulatory effect of DNA damage
on recombination. In vivo, a set of enzymes must be involved with
recA protein, including some that resolve the recombination intermediates. However, the economy with which the small recA protein
accomplishes both central steps of recombination, namely synapsis

38. Cox, M. M., and Lehman, I. R. (1981) PNAS 78, 6018; Kahn, R., Cunningham, R. P., DasGupta, C.,
and Radding, C. M. (1981) PNAS 78, 4786; West, S. C., Cassuto, E., and Howard-Flanders, P.
(1981) PNAS 78, 6149.
39. Cox, M. M., and Lehman, I. R. (1981) PNAS 78, 3433; DasGupta, C., and Radding, C. M. (1982)
PNAS 79, 762; Livneh, Z., and Lehman, I. R. (1982) PNAS 79, in press.

and heteroduplex formation, suggests a general mechanism based on a few principles.

## RecA Protein as a Proteolytic Inducer of SOS Functions

In addition to its role in homologous recombination, recA protein is required for the coordinate regulation of a set of cellular functions that occur in response to single-stranded DNA generated by DNA damage or interruption of DNA replication (Fig. S16-1). These functions include induction of prophages, enhanced mutagenesis, induction of long-patch excision-repair, inhibition of septation during cell division, colicin induction, increased expression of its own gene, and induction of uvrA, uvrB, umuC, and other damage-inducible (din) genes.

It is now clear that expression of these SOS functions is a consequence of the proteolytic cleavage by recA protein of bacterial repressors (e.g., lexA protein) that control other genes for the SOS functions. Thus, λ repressor is cleaved in vitro into two fragments by a ternary complex consisting of recA protein, single-stranded DNA, and ATP (or its analog ATPγS).[40] The lexA protein, a negative regulator, is the repressor of the recA gene and other genes involved in the SOS response, and it is also cleaved in vitro by a similar ternary complex.[41]

In vivo, the single-stranded DNA effectors that stimulate recA protein to proteolysis are probably gaps that result during excision-repair or from blockage of chromosomal replication at sites of DNA damage. Thus, the wide variety of DNA-damaging agents and inhibitors of DNA synthesis that induce the SOS response in E. coli may do so by creating or stabilizing single-stranded regions in the chromosome.[42]

An altered form of recA protein caused by the tif-1 mutation is active in vivo as an inducer of SOS functions in the absence of DNA damage. The mutant protein responds as a protease in vitro to short oligonucleotides that fail to stimulate the wild-type protein.[43] Thus, this altered recA protein may complex with short, single-stranded regions or gaps that normally occur near the growing fork of replicating chromosomes and are too short to activate the recA+ enzyme.

40. Craig, N. L., and Roberts, J. W. (1980) Nat. 283, 26.
41. Little, J. W., Mount, D. W., and Yanisch-Perron, C. R. (1981) PNAS 78, 4199; Brent, R., and Ptashne, M. (1981) PNAS 78, 4204.
42. Phizicky, E. M., and Roberts, J. W. (1981) Cell 25, 259; Weinstock, G. M., and McEntee, K. (1981) JBC 256, 10883.
43. McEntee, K., and Weinstock, G. M. (1981) PNAS 78, 6061.

The cellular abundance of recA protein disclosed by radioimmune assays[44] is 2000 to 5000 molecules per cell in the uninduced state, with increases up to 100-fold upon induction. In view of these high levels, the large stoichiometric quantities of recA protein needed for recombination in vitro and the huge amounts needed for protease activity in vitro no longer seem unphysiological.

The regulatory role of recA protein for the SOS state, while prominent, is not exclusive. Other genes and proteins (e.g., *recF*, *recBC*) participate in responding to different types of DNA damage.[45] Beyond the induction of genetic functions to repair DNA, the mutagenic responses in the SOS state may represent a drive to acquire increased genetic potential.[46] The significance of the SOS system for studies of animal cells, in which a comparable SOS state has yet to be demonstrated, relates to the great capacity of these cells to respond to DNA-damaging agents and to the possibility that carcinogenesis and altered genetic potential of germ cells may have a similar basis.[47]

Site-Specific Recombination[48]

Two types of recombination, distinct from general homologous recombination, include the conservative form exemplified by phage λ, in which DNA replication is not required, and the duplicative type typified by transposable elements like Tn3 and phage Mu.[49] Integration and excision of phage λ require a single breakage and joining event between specific DNA segments of the phage and host; transposition of Tn3 involves a replicative fusion event in which the donor replicon becomes incorporated in its target genome with a copy of the transposon at each end, followed by a site-specific resolution. The latter is conservative, as in phage λ integration, and yields an intact donor replicon still carrying the transposon and a new copy of the transposon at the target site (Fig. S16-4).

*Integration-excision by phage* λ.[50] Although formally a reversible DNA breakage and joining reaction, the integration-excision reaction for phage λ has directional specificity: The recombining sites in the phage and host genomes are distinct (even though both have a 15-bp common core), and the forward and reverse reactions are distinctive

44. Paoletti, C., Salles, B., and Giacomoni, P., personal communication.
45. McPartland, A., Green, L., and Echols, H. (1980) *Cell* 20, 731.
46. Echols, H. (1981) *Cell* 25, 1.
47. Echols, H. (1981) *Cell* 25, 1.
48. *Movable Genetic Elements* (1981) *CSHS* 45; Sherratt, D., Arthur, A., and Dyson, P. (1981) *Nat.* 294, 608.
49. Campbell, A. (1981) *CSHS* 45, 1.
50. Nash, H. A. (1981) *Ann. Rev. Gen.* 15, 143; *Enzymes* 14, 471.

and separately regulated. Int protein is needed for both directions, but Xis for excision only:[51]

$$\text{P o P}' + \text{B o B}' \underset{\text{Int} + \text{Xis}}{\overset{\text{Int}}{\rightleftharpoons}} \text{B o P}' + \text{P o B}'$$

| P o P′ | B o B′ | | B o P′ | P o B′ |
|--------|--------|--------|--------|--------|
| Phage site | Host site | Int + Xis | Left | Right |
| | | | Prophage ends | |

The reaction also requires two host-encoded and regulated proteins, designated HimA and HimD (or Hip).[52] Much of the complexity probably derives from the need of the virus to control the direction of the reaction, toward integration in lysogenization after infection, or toward excision during induction from the prophage state.

In vitro studies have clarified the reactive sites and components. As judged by DNase-protection experiments, the Int protein binds to four distinct sites in the phage attachment site, two in the left (P) region, one in the right (P′) region, and one in the common core (o) region.[53] Remarkably, the tightest binding site appears to be in the P′ region, centered some 55 bp from the right end of the core region.[54] The DNA strand exchange occurs in the core region, producing a staggered cleavage 7 bp apart.[55] The host protein has been purified as a dimer, one component of which is HimA.[56] The host integration factor is absolutely required for λ integrative recombination in vitro[57] and in vivo, but its role in host functions remains to be established. Because the purified integrase has type I topoisomerase activity (Section S9-11) and shows no ATP requirement, formation of a covalent enzyme-substrate, phosphodiester intermediate seems certain.[58]

*Duplicative fusion and resolution by Tn3*.[59]   For the transposable element Tn3, transposition occurs in two stages (Fig. S16-4). First, the transposon is replicated in a process that fuses the donor and recipient genomes into a *cointegrate* structure. The cointegrate has

51. Echols, H. (1980) in *The Molecular Genetics of Development*, Academic Press, New York, p. 1; Nash, H. A. (1981) *Ann. Rev. Gen. 15*, 143.
52. Miller, H. I., Kikuchi, A., Nash, H. A., Weisberg, R. A., and Friedman, D. I. (1978) *CSHS 43*, 1121; Miller, H. I., and Nash, H. A. (1981) *Nat. 290*, 523; Miller, H. I., Kirk, M., and Echols, H. (1981) *PNAS 78*, 6754.
53. Hsu, P.-L., Ross, W., and Landy, A. (1980) *Nat. 285*, 85.
54. Hsu, P.-L., Ross, W., and Landy, A. (1980) *Nat. 285*, 85; Davies, R. W., Schreier, P. H., Kotewicz, M. L., and Echols, H. (1979) *N. A. Res. 7*, 2255.
55. Mizuuchi, K., Weisberg, R., Enquist, L., Mizuuchi, M., Buraczynska, M., Foeller, C., Hsu, P.-L., Ross, W., and Landy, A. (1981) *CSHS 45*, 429.
56. Miller, H. I., Kikuchi, A., Nash, H. A., Weisberg, R. A., and Friedman, D. I. (1978) *CSHS 43*, 1121; Nash, H. A., and Robertson, C. A. (1981) *JBC 256*, 9246.
57. Nash, H. A., and Robertson, C. A. (1981) *JBC 256*, 9246.
58. Kikuchi, A., and Nash, H. A. (1979) *PNAS 76*, 3760.
59. Kleckner, N. (1981) *Ann. Rev. Gen. 15*, 341; Cohen, S. N., and Shapiro, J. A. (1980) *Sci. Amer. 242*, No. 2, 40; Bukhari, A. I. (1981) *TIBS 6*, 56.

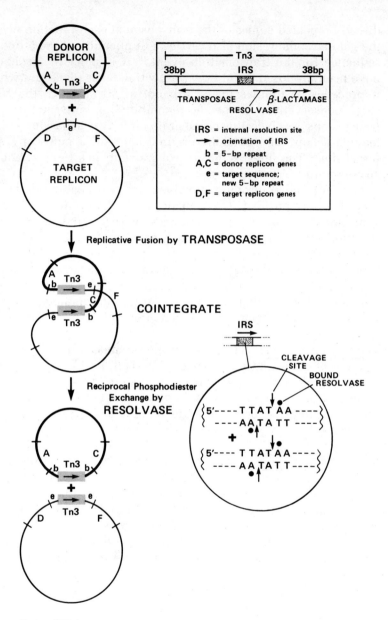

FIGURE S16-4
Scheme for replicative transposition of transposon, Tn3. See text for
details.

directly repeated copies of the transposon at each junction between the genomes; the fusion also carries a duplication of the 5-bp target sequence for the transposition event.[60] A number of mechanisms have been proposed,[61] but how replication takes place and whether it precedes or follows fusion are not at all clear. In the second stage, conservative site-specific recombination between duplicated transposons resolves the cointegrate into two separate replicons leaving a copy of Tn3 at its original site and a new copy at the target site.[62] The site-specific crossover occurs in a 19-bp region with 16 consecutive AT pairs.[63]

The Tn3-coded proteins involved in transposition have been identified and purified. The transposase is the product of the *tnpA* gene, and the resolution protein or resolvase (21 kdal) is specified by the *tnpR* gene.[64] The purified tnpA protein binds tightly to single-stranded DNA, but no other activities have so far been identified.[65] The resolvase, with $Mg^{2+}$, catalyzes site-specific recombination in vitro.[66] Staggered, single-strand cuts separated by two base pairs and a 5'-linked covalent enzyme-substrate intermediate are formed, reminiscent of topoisomerase activity (Section S9-12).

*Other transposons.* The recombination mechanisms for a large number of other transposons (e.g., Tn5, Tn9, Tn10) and IS (insertion sequence) elements are less well understood and may be different from that of the Tn3 family. Cointegrates do not appear to be intermediates; they are formed but are stable and an encoded resolvase is lacking.

*Transpositions in nature.*[67] First recognized as a complex genetic phenomenon in corn, the molecular traces of transpositional events are now seen in every reach of the bacterial and animal world (Fig. S16-5). Integration of DNA copies of RNA tumor virus genomes, numerous locations of transposable elements in yeast, *Drosophila*, bacteria, phage, and plasmids, all discovered in a short time, argue for an important role for transpositions in somatic and evolutionary change.

60. Shapiro, J. A. (1979) *PNAS 76*, 1933; Harshey, R. M., and Bukhari, A. I. (1981) *PNAS 78*, 1090; Galas, D. J., and Chandler, M. (1981) *PNAS 78*, 4858.
61. Arthur, A., and Sherratt, D. (1979) *MGG 175*, 267; Shapiro, J. A. (1979) *PNAS 76*, 1933.
62. Gill, R., Heffron, F., Dougan, G., and Falkow, S. (1978) *J. Bact. 136*, 742; Sherratt, D., Arthur, A., and Burke, H. (1981) *CSHS 45*, 275; Reed, R. R. (1981) *PNAS 78*, 3428.
63. Reed, R. R. (1981) *PNAS 78*, 3428.
64. Arthur, A., and Sherratt, D. (1979) *MGG 175*, 267; Sherratt, D., Arthur, A., and Burke, H. (1981) *CSHS 45*, 275; Reed, R. R. (1981) *PNAS 78*, 3428; Chou, J., Casadaban, M. J., Lemaux, P. G., and Cohen, S. N. (1979) *PNAS 76*, 4020; Gill, R., Heffron, F., and Falkow, S. (1979) *Nat. 282*, 797.
65. Fennewald, M. A., Gerrard, S. P., Chou, J., Casadaban, M. J., and Cozzarelli, N. R. (1981) *JBC 256*, 4687.
66. Reed, R. R. (1981) *Cell 25*, 713; Reed, R. R., and Grindley, N. D. F. (1981) *Cell 25*, 721.
67. Calos, M. P., and Miller, J. H. (1980) *Cell 20*, 579; Temin, H. M. (1980) *Cell 21*, 599.

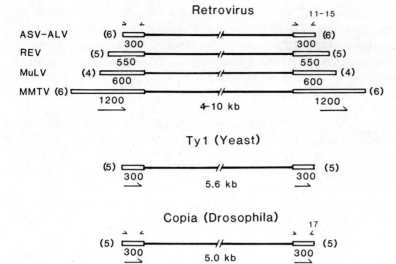

FIGURE S16-5
Transposable elements in nature represented by retroviruses, yeast,
*Drosophila*, and bacterial transposons. The arrowheads signify
inverted-repeat sequences. (Courtesy of Professor H. Hanafusa.)

# S16-5  Transformation[68]

Genetic engineering has put studies of transformation into high
gear. Once limited to a few bacterial species for which naked DNA
transfer is a natural and highly efficient process, there now appears
to be no experimental barrier for DNA uptake and assimilation by
the widest range of bacterial, plant, and animal cells. Advances in
this area make a section on transformation worthy of chapter status.

Natural Transformation Systems

*Gram-negative bacteria.*  *Hemophilus influenzae* and *H. parain-
fluenzae* are effective in transfer of homologous duplex DNA. Distributed through *Hemophilus* DNA at 4-kb intervals are 11-bp uptake
sites (5'-AAGTGCGGTCA)[69] recognized by several recipient-cell
surface receptor proteins. The assimilated duplex DNA, protected by

68.  Smith, H. O., Danner, D. B., and Deich, R. A. (1981) *ARB 50*, 41.
69.  Danner, D. B., Deich, R. A., Sisco, K. L., and Smith, H. O. (1980) *Gene 11*, 311.

a binding protein[70] or vesicle,[71] is then integrated into the genome as a single strand. Transformation in *Neisseria*, an organism of clinical importance, may employ a similar system.[72] Cloning of *Hemophilus* uptake sites and receptor-cell competence genes should make it possible to endow other bacteria (e.g., *E. coli*) with this remarkable natural transformation system.

*Gram-positive bacteria.* Possession of a multilayered peptidoglycan cell wall instead of the single layer of peptidoglycan and outer lipid membrane found in Gram-negative bacteria dictates a different DNA transfer system. (Note that the *E. coli* cell envelope in Fig. 13-3, p. 452, which shows peptidoglycan as multilayered, is misleading.) Transformation in *Streptococcus* (e.g., *S. pneumoniae*) and *Bacillus* (e.g., *B. subtilis*) differs from the *Hemophilus* system in: (i) isolatable competence factors that are protease sensitive, (ii) nonspecificity of DNA uptake, and (iii) single-stranded state of the DNA taken up.

Interaction of a competence factor with a cell surface receptor induces a competence-specific autolysin that exposes a membrane-associated DNA-binding protein[73] and nuclease. Endonucleolytic cleavage[74] of bound DNA is followed by uptake of one strand while the other is degraded.[75] The DNA strand organized by a competence-specific protein in an eclipse complex[76] is integrated by replacement of a genomic strand. Correction of mismatched bases may depend on methylation mechanisms to identify and repair the donor segment (Section S16-7).

*Transfection with viral and plasmid DNA.* Infection of a cell with naked viral or plasmid DNA, called transfection, employs the natural transformation systems but is $10^4$ times less effective than chromosomal DNA. A failure to recognize the single-stranded DNA form of the virus or plasmid may be responsible in Gram-positive bacteria; other reasons may explain the poor efficiencies in Gram-negatives.

Generalized Transformation Techniques

*Transformation of prokaryotes.* Modification of the $CaCl_2$ treatment[77] of recipient cells has raised the efficiency of artificial transformation to levels (20 percent of viable cells; $2 \times 10^7$ trans-

70. Sutrina, S. L., and Scocca, J. J. (1979) *J. Bact. 139*, 1021.
71. Kahn, M., Concino, M., Gromkova, R., and Goodgal, S. H. (1979) *BBRC 87*, 764.
72. Dougherty, T. J., Asmus, A., and Tomasz, A. (1979) *BBRC 86*, 97.
73. Ceglowski, P., Kawczynski, M., and Dobrzanski, W. T. (1980) *J. Bact. 141*, 1005.
74. Lacks, S. (1979) *J. Bact. 138*, 404; Rosenthal, A. L., and Lacks, S. A. (1980) *JMB 141*, 133.
75. Claverys, J. P., Roger, M., and Sicard, A. M. (1980) *MGG 178*, 191.
76. Morrison, D. A., and Baker, M. F. (1979) *Nat. 282*, 215; Morrison, D. A., and Mannarelli, B. (1979) *J. Bact. 140*, 655.
77. Dagert, M., and Ehrlich, S. D. (1979) *Gene 6*, 23; Chang, S., and Cohen, S. N. (1979) *MGG 168*, 111.

formants per microgram of plasmid DNA) that make this a general method for bacteria, *E. coli* included. Remarkably, the efficiencies for plasmid and viral DNA transformations are higher than for chromosomal DNA in bacteria that lack a natural transformation system, the opposite of what obtains in bacteria that have a natural system (see above). The mechanisms of the calcium effect and DNA uptake in the artificial system are still obscure beyond evidence that, contrary to natural systems, the donor DNA can be non-homologous (unlike Gram-negatives) and remain double-stranded during uptake (unlike Gram-positives).

*Transformation of animal cells.*[78] DNA fragments in the 20-kb range, subchromosomal segments, and even intact metaphase chromosomes can be transferred to mammalian cells. Transformation by naked DNA, commonly as a calcium phosphate precipitate, is an extension of transfection studies with oncogenic viral DNA and infection by virions that introduce selective markers, such as a functional thymidine kinase or methotrexate resistance. While many cells take up DNA, very few ($10^{-5}$ to $10^{-6}$) become stably transformed. Mechanisms of DNA entry (possibly by phagocytosis), its migration to the nucleus, and the subsequent fate of the DNA are known only in outline.[79] Either a reconstitution of the DNA into a large molecule with its own centromeric region or an association with a host chromosome represents alternative means for producing stable transformants. A novel, potentially general method of introducing a DNA segment at chromosomal locations, demonstrated in *Drosophila*, is based on incorporating the segment in a transposon that inserts itself in the chromosome.[79a]

## S16-7  Modification by Methylation[80]

Site-specific modification, as distinguished from overall DNA methylation and other modifications,[81] can be examined more readily with improved techniques, and the question of its biological role, particularly in eukaryotes, is intriguing.

### Prokaryotes

Aside from methylations that serve in the restriction-modification system (p. 641), only one other function for DNA methylation in

78. Klobutcher, L. A., and Ruddle, F. H. (1981) *ARB 50,* 533.
79. Pellicer, A., Robins, D., Wold, B., Sweet, R., Jackson, J., Lowy, I., Roberts, J. M., Sim, G. K., Silverstein, S., and Axel, R. (1980) *Science 209,* 1414.
79a. Rubin, G. M., and Spradling, A. C., personal communication.
80. Hattman, S. E. (1981) *Enzymes 14,* 517.
81. Warren, R. A. J. (1980) *Ann. Rev. Microb. 34,* 137.

prokaryotes appears to be clarified. In the course of replication, the unmethylated nascent strand can be distinguished from the methylated parental strand as the one in need of repair should a mismatched base have been incorporated.[82] Higher mutation rates in *E. coli* mutants deficient in adenine methylase (*dam*)[82a] may be explained by the lack of this function. By contrast, mutant cells that lack the major cytosine methylase show no phenotypic defects.

In phage Mu DNA, adenine residues are modified to a novel form[83] to the extraordinary level of 15 percent. Expression of a Mu gene (*mom*) and the host *dam* are needed. The purpose of this modification may be to protect Mu DNA against restriction.

Eukaryotes[84]

Methylation of DNA in eukaryotes differs from the pattern in prokaryotes in these features:

(i)   The principal modified base is 5-methylcytosine except in some unicellular forms which have mostly methyladenine. 5-Methylcytosine is the only modified base in mammalian DNA; 90 percent is in the symmetrical sequence, CpG.

(ii)   The extent of methylation of available residues is only partial; only 50 to 70 percent of mammalian CpG sequences are modified, depending on the species and tissue.

(iii)   Viral DNA is not methylated.

(iv)   For lack of known restriction-modification systems in eukaryotes, methylation cannot be assigned this function.

Methylation of mammalian DNA is of current interest as a possible device for gene regulation and cellular differentiation. The correlations between hypomethylation and gene activation or conversely between extensive methylation and transcriptional inactivity[85] are impressive, but the causal relationships have not been proved. In any case, the apparent absence of methyl groups in *Drosophila* DNA suggests that in these highly differentiated creatures at least, an alternative to methylation must exist.

Several aspects of mammalian DNA methylation suggest it as a mechanism for gene regulation.

82. Wagner, R., Jr., and Meselson, M. (1976) *PNAS 73*, 4135; Radman, M., Wagner, R. E., Jr., Glickman, B. W., and Meselson, M. (1980) *Prog. Environ. Mutagen.* (Alacevic, M., ed.) Elsevier/North Holland Biomedical Press, Amsterdam, p. 121.
82a. Herman, G. E., and Modrich, P. (1982) *JBC 257*, 2605.
83. Hattman, S. E. (1979) *J. Virol. 32*, 468.
84. Razin, A., and Riggs, A. D. (1980) *Science 210*, 604; Wigler, M. H. (1981) *Cell 24*, 285; Felsenfeld, G., and McGhee, J. (1982) *Nat. 296*, 602.
85. Sobieski, D. A., and Eden, F. C. (1981) *N. A. Res. 9*, 6001; Naveh-Many, T., and Cedar, H. (1981) *PNAS 78*, 4246.

(i) Methylation can alter the state of DNA and depress the affinity of a repressor for an operator site.[86]

(ii) Methylation is symmetrical in both DNA strands. Assume that methylation follows replication and that a methyl transferase exists that only acts on a CpG in the newly synthesized strand when the CpG opposite in the template strand is already methylated. As a consequence, the unmethylated sites in the parent will remain unmethylated in the progeny, just as sites on hemimethylated DNA signal methylation of specific sites on the progeny DNA. It follows that methylation patterns are somatically (clonally) heritable.[87]

(iii) Methylation patterns differ between tissues.[88] The hyper-methylated state of germ or undifferentiated stem cells can, with hypomethylation during replicative stages, lead to progressive demethylation.

(iv) 5-Azacytidine, an analog of cytidine, causes fibroblasts to differentiate into muscle cells, presumably by inhibiting DNA methylation.[89] Reversal of X-chromosome inactivation by 5-azacytidine is another dramatic effect that suggests DNA methylation as the basis for stable inactivation of X-chromosome genes during normal mammalian differentiation.[90] Should these interpretations of 5-azacytidine action prove valid, then the unduly high activity of a cellular oncogene (Section S13-6) can in some instances be due to demethylation rather than a mutation, a chromosomal rearrangement, or a viral integration.[91]

---

86. Fisher, E. F., and Caruthers, M. H. (1979) N. A. Res. 7, 401.

87. Stein, R., Gruenbaum, Y., Pollack, Y., Razin, A., and Cedar, H. (1982) PNAS 79, 61.

88. Sobieski, D. A., and Eden, F. C. (1981) N. A. Res. 9, 6001; Naveh-Many, T., and Cedar, H. (1981) PNAS 78, 4246.

89. Jones, P. A., and Taylor, S. M. (1980) Cell 20, 85; N. A. Res. 9, 2933; Creusot, F., Acs, G., and Christman, J. K. (1982) JBC 257, 2041.

90. Liskay, R. M., and Evans, R. J. (1980) PNAS 77, 4895; Mohandas, T., Sparkes, R. S., and Shapiro, L. J. (1981) Science 211, 393.

91. Hayward, W. S., Neel, B. G., and Astrin, S. M. (1981) Nat. 290, 475; Lapeyre, J.-N., and Becker, F. F. (1979) BBRC 87, 698.

# S17

# Synthesis of Genes and Chromosomes

## Abstract

Advances in the subjects covered in the preceding chapters have made some topics worthy of far more attention than accorded them in the current edition of *DNA Replication*. Gene and chromosome synthesis is an extreme case. The avalanche of new information has already created an annual series of monographs, many new scientific journals, and even some trade newspapers. Little choice is left but to limit the news coverage in this chapter to captions and references for the guidance of the nonspecialist.

Advances in gene and chromosome synthesis in the past two years have been marked by many demonstrations of the latent power of the technologies of analysis, synthesis, and recombination of DNA. A sequence of 1000 nucleotides can be analyzed in a day. The sequence of the 40-kb genome of T7 phage has been completely determined. Equally remarkable is the accuracy of nucleotide sequence analysis attested by comparisons of results from different laboratories. Automated chemical synthesis on solid supports enables a block of fourteen nucleotides to be made in a day. With trimers as building blocks, longer sequences are readily assembled and many of these can in turn be annealed and joined by DNA ligase. A 514-bp duplex containing the human leukocyte interferon gene, synthesized by ligation of 66 oligonucleotides, has been cloned in *E. coli.*

With the capacity to alter the sequence at every residue, the effects on genetic regulation and function can be assessed, and these will have major significance for understanding basic mechanisms and for application to clinical problems. Techniques for dissecting and rearranging chromosomes and for delivering DNA into virtually any cell have kept pace with analysis and chemical synthesis. Once in the recipient cell, recombinant DNA is replicated, expressed, and integrated into the genome. Genetic engineering has opened up enormous opportunities for studies of cellular function, architecture, and development. Amid all the excitement, the importance of enzymes as crucial reagents for analysis, synthesis, and recombination deserves special homage.

## S17-1 Origin of DNA on Earth

Leaving aside the proposal that life originated on another planet and was planted on Earth by rockets (directed panspermia),[1] and the controversy as to whether prebiotic chemistry on Earth was conditioned by a reducing atmosphere, a view seriously challenged by geologists, there must still be earnest interest in nonenzymatic mechanisms of RNA and DNA synthesis. Wherever in the universe template-directed synthesis of polynucleotides did take place, it is helpful to know the important catalytic role that metals, such as $Zn^{2+}$ and $Pb^{2+}$, can play.

Zinc is of special interest because it is an essential component of all DNA and RNA polymerases examined. With a poly C template, $Zn^{2+}$ alone can catalyze the assembly of an activated GMP derivative (guanosine 5′-phosphoimidazolide) into poly G chains of 30 to 40 residues in the natural 3′-5′ linkage.[2] This regiospecificity is remarkable in view of the far greater reactivity of the 2′-OH compared with the 3′-OH. Zinc action is likely directed at changing the geometry of the polynucleotide helix rather than activating the 3′-OH group. Compared to GMP, the other nucleotides (AMP, UMP, and CMP) are selected against by a factor of 200. Although $Pb^{2+}$ is an efficient catalyst,[3] the oligomers it produces are 2′-5′ linked, and the fidelity to template direction is far less than with $Zn^{2+}$.

1. Crick, F. (1981) *Life Itself, Its Origin and Nature,* Simon and Schuster, New York.
2. Lohrmann, R., Bridson, P. K., and Orgel, L. E. (1980) *Science 208,* 1464; Bridson, P. K., and Orgel, L. E. (1980) *JMB 144,* 567.
3. Lohrmann, R., and Orgel, L. E. (1980) *JMB 142,* 555; van Roode, J. H. G., and Orgel, L. E. (1980) *JMB 144,* 579.

Prebiotic synthesis of polynucleotides may have been influenced by interaction with short peptides. This possibility is supported by the effects that rather minor changes in the imidazole substituent have on efficiency and regiospecificity of poly G synthesis, even in the absence of $Zn^{2+}$. For example, the 2-methylimidazole GMP derivative generates 89 percent of oligomers, four or more residues long, connected predominantly by 3'-5' diester bonds in the absence of $Zn^{2+}$.[4]

## S17-2   Determination of DNA Sequence

The power of DNA sequencing technology continues to expand and produce novel insights into structure that ramify into many aspects of genetic evolution and function. As examples, analyses of mitochondrial DNAs reveal a new genetic code, anomalous tRNAs, additional codon-anticodon interactions, and a different initiation mechanism for protein synthesis.[5] Comparison of the germ line and somatic cell sequences of the immunoglobulins identifies somatic recombination and mutations that help account for the structures and diversity of antibodies.[6] The amino acid sequence of a protein, inferred from an open reading frame in the DNA sequence, can guide the chemical synthesis of oligopeptides, which can then serve as immunogens to elicit antibodies; these in turn can identify and isolate a previously unknown protein from a cell or tissue.[7]

The remarkable accuracy of DNA sequencing rests on the cross-checking afforded by analysis of both strands of the duplex, on the availability of numerous restriction enzymes that allow determination of multiple overlapping of tracts of sequence, and on the coding sense of cistronic regions. Both the chemical and the enzymatic replication methods for sequence analysis (Section 17-2) have been assisted by modifications in gel electrophoretic separations[8] that now resolve up to 500 residues and permit the analysis of 1000 nucleotides a day. Future improvements in gel reading and data entry,[9] and computer programs that assemble the final sequence,[10] should further increase the speed and extent of analysis.

4. Inoue, T., and Orgel, L. E. (1981) *J. Am. Chem. Soc. 103*, 7666.

5. Sanger, F. (1981) *Science 214*, 1205.

6. Tonegawa, S. (1979–80) Harvey Lectures *75*, 61; Ravetch, J. V., Siebenlist, U., Korsmeyer, S., Waldmann, T., and Leder, P. (1981) *Cell 27*, 583; Kim, S., Davis, M., Sinn, E., Patten, P., and Hood, L. (1981) *Cell 27*, 573.

7. Lerner, R. A., Sutcliffe, J. G., and Shinnick, T. M. (1981) *Cell 23*, 309.

8. Sanger, F., and Coulson, A. R. (1978) *FEBS Lett. 87*, 107.

9. Gingeras, T. R., and Roberts, R. J. (1980) *Science 209*, 1322.

10. Gingeras, T. R., Milazzo, J. P., Sciaky, D., and Roberts, R. J. (1979) *N. A. Res. 7*, 529; Staden, R. (1980) *N. A. Res. 8*, 3673; Special issue (1982) *N. A. Res. 10*, 1–456.

The chemical sequencing method is advantageous for the ease with which it can be set in motion and its ready applicability to most any DNA fragment.[11] The dideoxy method (now used exclusively among the enzymatic methods) once in place is effective for determining extensive, genome-length sequences. The requirement for single-stranded DNA, a limitation of enzymatic methods, has been circumvented in two ways. One is by degradation of a duplex fragment by exonuclease III ($3' \rightarrow 5'$; p. 324) or phage $\lambda$ exonuclease ($5' \rightarrow 3'$; p. 327) to generate long stretches of single-stranded template, ideal for primed-synthesis sequence analysis.[12] The other approach, termed "shotgun" DNA sequencing, is to clone random fragments 200–400 nucleotides long (produced by pancreatic DNase I in the presence of $Mn^{2+}$) in a single-stranded phage vector such as M13.[13] A $\beta$-galactosidase gene with an EcoRI restriction site is included in the M13 genome and provides a convenient site for insertion of the fragments and the means for identifying recombinant phage with the use of a suitable indicator system.

A collection of phage clones, each with a random fragment, encompasses the entire genome to be sequenced; a chemically synthesized oligonucleotide, complementary to the sequence flanking the cloning site, serves as a universal primer on single-stranded DNA extracted from the phage. Once again, simple cloning replaces the laborious and inexact fractionations of restriction digests. The only serious limitation of the shotgun method is that as the sequencing approaches completion, new information diminishes asymptotically; analysis of twenty random clones, with an average insert of 227 residues, yielded 3 kb of a 4.3-kb sequence, but analysis of another twenty to thirty clones was needed to complete the sequence.[14]

The sequences of unexplored chromosomal regions of special interest have been obtained by "walking" from nearby regions. With clones containing overlapping sequences, the 100-kb bithorax region of *Drosophila* that determines its segmental patterns has been reached after a jaunt of 300 or so kilobases.[15] Analyses of the sequences define transcriptional boundaries and show that most of the spontaneous mutations in this vast locus are the result of transpositional events.

11. Gilbert, W. (1982) *Science 214,* 1305.
12. Smith, A. J. H. (1979) *N. A. Res. 6,* 831; Zain, B. S., and Roberts, R. J. (1979) *JMB 131,*341; Guo, L.-H., and Wu, R. (1982) *N. A. Res. 10,* 2065.
13. Sanger, F., Coulson, A. R., Barrell, B. G., Smith, A. J. H., and Roe, B. A. (1980) *JMB 143,* 161; Anderson, S., Gait, M. J., Mayol, L., and Young, I. G. (1980) *N. A. Res. 8,* 1731; Messing, J., Crea, R., and Seeburg, P. H. (1981) *N. A. Res. 9,* 309.
14. Anderson, S. (1981) *N. A. Res. 9,* 3015.
15. Bender, W., Spierer, P., and Hogness, D. S. (1979) *J. Supramol. Struc.,* Supp. 3, p. 32.

## S17-3 Chemical Synthesis of Oligodeoxyribonucleotides

## S17-4 Chemical Synthesis of Genes

S227

SECTION S17-4:
Chemical Synthesis
of Genes

The forward leap in chemical synthesis of DNA rivals that of DNA sequencing. Synthesis of a 14-nucleotide sequence that previously consumed a year and could be undertaken by only a few laboratories in the world can now be done in a day by a trained technician most anywhere. Combination of an automated synthetic procedure that adds a nucleotide in a 30-minute cycle, practical procedures for isolation and characterization of the product, and reliable methods for enzymatic repair and ligation of the oligonucleotides produces a gene of 500 or more base pairs long on a time scale comparable with its isolation from natural sources.

### Oligodeoxyribonucleotides

Beyond the use of oligodeoxyribonucleotides as the building blocks of genes, their value is impressive in many other ways: as binding sites for a variety of regulatory and catalytic proteins; as linkers in splicing genes into chromosomes; as probes to isolate a genetic sequence from cells or tissues; as primers to synthesize or sequence DNA (or RNA); and as the source material for x-ray crystallography and other refined studies of DNA structure.

Synthesis on a solid-phase support of cellulose,[16] silica gel,[17] or resins of polydimethylacrylamide[18] or polystyrene,[19] rather than in solution, has become the method of choice because of speed, efficiency, and capacity for automation. In each cycle of nucleotide addition, unreacted reagents are rapidly removed and an unreacted 5'-hydroxyl group can be masked against further additions. The choice of suitably protected mono-, di-, or trinucleotides (p. 667) as building blocks is dictated by several considerations. Only 4 types of mononucleotides are needed, but 16 dimers and 64 trimers are required. However, the trimers are readily synthesized from monomers and dimers and can be stored safely.

16. Crea, R., and Horn, T. (1980) N. A. Res. 8, 2331.
17. Alvarado-Urbina, G., Sathe, G. M., Liu, W.-C., Gillen, M. F., Duck, P. D., Bender, R., and Ogilvie, K. K. (1981) Science 214, 270 Matteucci, M. D., and Caruthers, M. H. (1981) J. Am. Chem. Soc. 103, 3185.
18. Markham, A. F., Edge, M. D., Atkinson, T. C., Greene, A. R., Heathcliffe, G.R., Newton, C. R., and Scanlon, D. (1980) N. A. Res. 8, 5193.
19. Miyoshi, K., Arentzen, R., Huang, T., and Itakura, K. (1980) N. A. Res. 8, 5507.

The use of trimers not only saves time and effort in synthesizing long chains but also makes the separation of product from impurities more decisive because of the larger differences in size and charge. Oligomers of 37 and 34 nucleotides, synthesized by successive addition of trimers, were annealed to generate a duplex with overlapping 5' ends.[20] A flush-ended 55-bp duplex (containing the protein n' recognition site at the origin of phage $\phi$X174 DNA replication (Section S11-9)) was formed by filling in the ends with DNA polymerase I.[20]

Current procedures use the phosphite and phosphotriester methods with phosphorylating reagents such as methyl phosphodichloridites or aryl phosphodichloridates and their triazolide derivatives.[21,22] Others find advantage in using condensing reagents such as tri-isopropylbenzenesulphonyl tetrazolide or mesitylenesulphonyl 3-nitrotriazolide to form the internucleotide linkage.[23]

## Genes

Synthesis of the genes for small hormones, such as somatostatin (p. 670), insulin, and thymosin, was carried out in solution. With solid-phase support, the total synthesis of a 514-bp duplex containing the human leukocyte interferon gene has been achieved.[23] Engineered into the duplex are initiation and termination signals for expression in E. coli and appropriate restriction enzyme sites for insertion into plasmid vectors. The strategy for synthesis was based on the preparation of 66 duplex oligonucleotides ranging in size from 14 to 21 residues and designed with overlapping ends for use in enzymatic ligation after annealing of complementary fragments.

The native sequence of the interferon gene contains many repeated heptamers and octamers, all of which may be troublesome in ligation of synthetic fragments. These potential difficulties were taken into account, as were the preferred codon usages in E. coli, by making 71 substitutions within the constraints of the genetic code. E. coli clones containing the anticipated nucleotide sequence were obtained. Synthesis of active interferon can be expected in view of the enormous quantities of the protein produced by E. coli containing plasmids with cloned natural interferon genes.

The most impressive advantage of chemical synthesis over iso-

20. Crea, R., Vasser, M.P., and Struble, M. E., personal communication.
21. Alvarado-Urbina, G., Sathe, G. M., Liu, W.-C., Gillen, M. F., Duck, P. D., Bender, R., and Ogilvie, K. K. (1981) *Science 214*, 270; Matteucci, M. D., and Caruthers, M. H. (1981) *J. Am. Chem. Soc. 103*, 3185.
22. Broka, C., Hozumi, T., Arentzen, R., and Itakura, K. (1980) *N. A. Res. 8*, 5461.
23. Edge, M. D., Greene, A. R., Heathcliffe, G. R., Meacock, P. A., Schuch, W., Scanlon, D. B., Atkinson, T. C., Newton, C. R., and Markham, A. F. (1981) *Nat. 292*, 756.

lation of natural genes (or the cDNA copies of processed mRNAs) is the ease and precision of altering every residue in the chain. The striking effects of even a single base substitution in the regulation of gene expression by attenuation (Section S7-6),[24] let alone the profound influence of an amino acid change on the activity of a protein, demonstrate the power of creating analogs for basic studies and clinical applications.

## S17-5  Assembling Genes into Chromosomes: Recombinant DNA[25-27]

Developments in a major, worldwide research industry reported in daily newspapers, in monographs, and in every issue of twenty or more journals can hardly be summarized here in a page or two. In effect, this entire Supplement is based on the improved cloning devices, superior techniques for transforming bacterial, plant, and animal cells with DNA, and advanced techniques of DNA sequencing and chemical synthesis that have opened up vast opportunities for exploring gene evolution, structure, and function, and virtually every facet of cellular architecture and development. Only a few of the recent advances in genetic engineering and related technologies with broad practical applications need be cited as illustrative examples.

(i)   Creation of shuttle vectors (chimeric plasmids) that replicate autonomously and support gene expression in both a bacterial and a yeast cell or in a bacterial and an animal cell or in two different bacterial species (e.g., E. coli and B. subtilis).

(ii)   Production of vectors (i.e., plasmids), with mutations in copy-number control, that can be induced to levels of 1000 per cell and thus exceed the amount of chromosomal DNA; the level of a protein encoded by a gene in the "runaway" plasmid can be amplified 500-fold.

(iii)   Engineering of superior promoters of gene expression: genes under control of the trp or lac promoters can be induced or repressed by nutrients or analogs in the medium; transcriptional promoters as several tandem copies or in special locations can far exceed the capacities of other cell promoters. Similarly, superior ribosomal binding sites may be fashioned to enhance translation.

24. Yanofsky, C. (1981) Nat. 289, 751.
25. Recombinant DNA, Science 209, pp. 1317–1438 (1980).
26. Genetic Engineering (Setlow, J. K., and Hollaender, A., eds.) vol. 1–3 (1979–1981) Plenum Press, New York.
27. Berg, P. (1981) Science 213, 296.

(iv)  Exploitation of the permissiveness of a bacterium (e.g., *E. coli*) in responding to a strongly promoted gene: protein overproduction can reach bizarre levels; human insulin, *E. coli* recA protein, or *E. coli* tryptophan synthetase can make up 50 percent or more of the cellular protein. Perhaps the bacterium, with its proteolytic disposal mechanisms overwhelmed, can sequester a huge deposit of a foreign protein (or one of its own proteins normally present in a trace amount), so that a simple cell lysate yields the protein in a nearly pure form.

(v)  Fusion of the DNA sequence for the signal peptide of a membrane or secretory protein to the gene for a cytoplasmic protein can translocate the latter to a membranous or extracellular location.

(vi)  Introduction of a centromeric sequence into a yeast plasmid endows it with the status of a stable, functional yeast chromosome.

(vii)  Synthesis of cDNAs by refined use of enzymes and linkers enables libraries to be constructed representing celluar mRNAs, including trace varieties, in their entirety and with high fidelity.

## S17-6   Homage to Enzymes

Enzymes, including those that make recombinant DNAs, have not received the attention they deserve. The ease of analyzing existing DNA and creating new DNA and new species has diverted the spotlight from the more laborious isolation and characterization of enzymes. Our want of knowledge about the enzymes that redesign the genes and chromosomes as well as act out those designs in cellular form and function needs to be redressed. In examining the cast of enzymes that is largely responsible for the drama of genetic engineering (Table S17-1), it is notable that few of the performers were discovered in order to act these roles.

TABLE S17-1
Enzymatic tools of genetic engineering

| Class | Enzyme[a] | Class | Enzyme[a] |
|---|---|---|---|
| Nucleases | restriction endonucleases<br>S1 endonuclease<br>exonuclease III<br>λ exonuclease<br>*Bal*31 exonuclease | Polymerases | DNA polymerase I<br>T4 DNA polymerase<br>reverse transcriptase<br>terminal deoxynucleotidyl transferase |
| Ligases | T4 DNA ligase<br>T4 RNA ligase | Phosphatases | *E. coli* alkaline phosphatase<br>calf alkaline phosphatase |
| Kinases | T4 polynucleotide kinase<br>DNA kinase | | |

[a]For details in this Supplement and in *DNA Replication,* consult the indexes and Singer, M. (1979) in *Genetic Engineering* (Setlow, J. K., and Hollaender, A., eds.), Plenum Press, New York, p. 1.

# Publication Abbreviations

| | |
|---|---|
| *Adv. Virus Res.* | Advances in Virus Research |
| *Ann. Rev. Gen.* | Annual Review of Genetics |
| *Ann. Rev. Microb.* | Annual Review of Microbiology |
| *ARB* | Annual Review of Biochemistry |
| *B.* | Biochemistry |
| *Bact. Rev.* | Bacteriological Reviews: Microbiological Reviews |
| *BBA* | Biochimica et Biophysica Acta |
| *BBRC* | Biochemical and Biophysical Research Communications |
| *BJ* | The Biochemical Journal |
| *Comp. Virol.* | Comprehensive Virology |
| *Crit. Rev. Bioch.* | CRC Critical Reviews of Biochemistry |
| *CSHL* | Cold Spring Harbor Laboratories, New York |
| *CSHS* | Cold Spring Harbor Symposia on Quantitative Biology |
| *EJB* | European Journal of Biochemistry |
| *Enzymes* | *The Enzymes,* 3rd ed. (Boyer, P. D., ed.) (1981), Academic Press, New York |
| *FEBS Lett.* | Federation of European Biochemical Societies Letters |
| *ICN–UCLA Symposia* | ICN–UCLA Symposia on Molecular and Cellular Biology, Academic Press, New York |
| | Vol. 19 (1980): *Mechanistic Studies of DNA Replication and Genetic Recombination* (Alberts, B., ed.) |
| | Vol. 20 (1981): *Structure and DNA-Protein Interactions of Replication Origins* (Ray, D. S., and Fox, C. F., eds.) |
| | Vol. 22 (1981): *The Initiation of DNA Replication* (Ray, D. S., ed.) |
| *JBC* | Journal of Biological Chemistry |
| *JMB* | Journal of Molecular Biology |
| *J. Bact.* | Journal of Bacteriology |
| *J. Gen. Virol.* | Journal of General Virology |
| *J. Virol.* | Journal of Virology |
| *MGG* | Molecular and General Genetics |
| *Mol. Biol. Plasmids* | *Molecular Biology, Pathogenicity and Ecology of Bacterial Plasmids* (Levy, S., Clowes, R. and Koenig, E., eds.) (1981), Plenum Press, New York |
| *Nat.* | Nature |
| *N. A. Res.* | Nucleic Acids Research |
| *PNAS* | Proceedings of the National Academy of Sciences, U.S.A. |
| *Prog. N. A. Res.* | Progress in Nucleic Acid Research and Molecular Biology |
| *TIBS* | Trends in Biochemical Sciences |
| *Virol.* | Virology |

# Structures

| | |
|---|---|
| ADP, etc | 5'-(pyro)-diphosphates of adenosine, etc. |
| *amber* mutant | nonsense mutant in which the codon has been altered to the chain-terminating UAG |
| AMP, GMP, IMP, UMP, CMP | 5'-phosphates of ribonucleosides of adenine, guanine, hypoxanthine, uracil, cytosine |
| 2'-AMP, 3'-AMP (5'-AMP), etc. | 2'-, 3'- (and 5'-, where needed for contrast) phosphates of the nucleosides |
| Ap, etc. | 3'-AMP, etc. |
| AT, GC | base pairs of adenine with thymine, and of guanine with cytosine, or their respective deoxyribonucleotides |
| ATP, etc. | 5'-(pyro)-triphosphates of adenosine, etc. |
| $^{14}$C | a radioactive isotope of carbon with a half-life of 5730 years |
| cAMP | 3', 5'-cyclic AMP |
| capsid | a coat protein unit of a virus |
| cDNA | complementary DNA obtained by (reverse) transcription of RNA |
| concatemer | a chain of an unstated number of repeated units of duplex DNA |
| d | 2'-deoxyribo (nucleoside, nucleotide) |
| $(dA)_{4000}$ | homopolymer chain of about 4000 deoxyriboadenylate residues |
| dAMP, etc. | 5'-phosphates of 2'-deoxyribosyladenine, etc. |
| d-A$_p$G | deoxyadenosine linked 3' to 5'-phosphodeoxyguanylate (see Fig. 1-2) |
| D-loop | a DNA replicative intermediate in which only one chain of a duplex region is replicated, leaving the other single strand displaced (see Fig. 11-3) |
| $dna_{ts}$ | genetic locus for a replication protein in a temperature-sensitive (conditionally lethal) mutant |
| dNMP | 5'-phosphate of a 2'-deoxyribonucleoside, in general |
| $^3$H | tritium, a radioactive isotope of hydrogen with a half-life of 12.26 years |
| holoenzyme | complete enzyme including cofactors, small or large |
| NEM | N-ethylmaleimide |
| NTP | nucleoside triphosphate |
| oligo dA | homopolymer chain of deoxyriboadenylate of undefined length, generally under several hundred residues |
| operon | adjacent genes regulated as a group by an operator sequence and its repressor. |
| P or p | phosphate (in compounds) |
| $^{32}$P | a radioactive isotope of phosphorus with a half-life of 14.3 days |
| pA, pG, pI, pU, pC | AMP, GMP, IMP, UMP, CMP |
| P$_i$, PP$_i$ | orthophosphate, inorganic pyrophosphate |

| phosphodiester | $$R\!-\!O\!-\!\overset{\displaystyle O}{\underset{\displaystyle O^-}{\overset{\|}{\underset{\|}{P}}}}\!-\!O\!-\!R$$ |
|---|---|
| poly dA, etc. | homopolymer chain of deoxyriboadenylate of undefined length, generally over 1000 residues |
| poly dA · poly dT | homopolymer chains of poly dA and poly dT associated by base pairing |
| poly d(AT) | copolymer chain of alternating A and T residues associated with another such chain by base pairing |
| poly rA · poly rU | homopolymer chains of adenylate and uridylate associated by base pairing |
| processive | repetitive enzyme action without dissociation between steps; nondistributive |
| prophage | provirus stage of a temperate phage |
| provirus | state of a virus, integrated into the host cell chromosome |
| PRPP | 5-phosphoribosylpyrophosphate |
| r | ribo (nucleoside, nucleotide) |
| retrovirus | RNA virus possessing a reverse transcriptase |
| RF | double-stranded, circular, replicative form of DNA |
| RFI | covalently closed RF |
| RFII | RF with discontinuity in at least one strand |
| rDNA | DNA encoding ribosomal RNA |
| $\sigma$ | sigma subunit of RNA polymerase; also the lariat-shaped DNA replicative intermediate in a rolling-circle mechanism |
| SS | single-stranded, circular DNA |
| suppressor mutation | mutation that restores a function lost by another mutation at another genetic site |
| $\theta$ | theta structure, a DNA replicative intermediate (see Fig. 11-3) |
| vegetative phage | free state of a phage |
| virion | a free virus particle |

# Quantities

| | | | |
|---|---|---|---|
| m | milli | $(10^{-3})$ | |
| $\mu$ | micro | $(10^{-6})$ | |
| n | nano | $(10^{-9})$ | |
| p | pico | $(10^{-12})$ | |
| M | molar | | (mole/liter) |
| mM | millimolar | $(10^{-3}$ M) | (mmole/liter) |
| $\mu$M | micromolar | $(10^{-6}$ M) | ($\mu$mole/liter) |
| nM | nanomolar | $(10^{-9}$ M) | (nmole/liter) |
| cm | centimeter | $(10^{-2}$ meter) | |
| mm | millimeter | $(10^{-3}$ meter) | |
| $\mu$m | micrometer | $(10^{-6}$ meter) | (micron) |
| nm | nanometer | $(10^{-9}$ meter) | (millimicron, m$\mu$) |
| Å | angstrom | $(10^{-10}$ meter) | (0.1 nm) |
| ml | milliliter | | |
| $\mu$l | microliter | | |
| g | gram | | |
| mg | milligram | | |
| $\mu$g | microgram | | |
| cpm | counts per minute of radioactivity | | |
| Ci | Curie | | |
| S | Svedberg unit of sedimentation | | |
| ° | degree Celsius | | |
| $K_m$ | Michaelis constant | | |
| bp | base pairs | | |
| kb | number of base pairs, in thousands; length of DNA, in $\mu$m $\times$ 1/3 | | |
| $T_m$ | thermal midpoint of melting, °C | | |
| $V_{max}$ | maximal enzyme rate | | |
| dalton | unit of mass equal to one-twelfth the mass of an atom of carbon-12 | | |
| kdal | kilodaltons; one thousand daltons | | |
| Mdal | megadaltons | | |
| mol. wt. | molecular weight, expressed in grams | | |

# Author Index

# Subject Index

(The Subject Index gives references both to *DNA Replication* and to this Supplement. Definitions and formulas are noted in **BOLD** type.)

934H